Applications of Calculus to
Biology and Medicine
Case Studies from Lake Victoria

Applications of Calculus to
Biology and Medicine
Case Studies from Lake Victoria

Nathan C Ryan
Bucknell University, USA

Dorothy Wallace
Dartmouth College, USA

NEW JERSEY · LONDON · SINGAPORE · BEIJING · SHANGHAI · HONG KONG · TAIPEI · CHENNAI · TOKYO

Preface

This book is about how quantitative reasoning plays out in simple biological examples. The quantitative reasoning the reader will engage in does not end with discussions of units and error margins — to us it is at its most useful when thoroughly integrated with traditional mathematical content such as calculus. The book, as we've conceived it, is meant to highlight the importance of calculus and other more advanced mathematical topics in the biological sciences in general.

Mathematics in the Biological Sciences

It is more important than ever that researchers and practitioners in biological fields know how to think quantitatively and use mathematical tools to their advantage. One can get a feel for the growth of mathematical tools in biology by comparing research articles found by Google Scholar, as in the table below.

On May 31, 2017 a brief Google Scholar search revealed the following number of items under these categories. Two searches were done on six terms, one for the whole database and the other for entries since 2016. In the percentages given below, we assume that the items found with the modifier "mathematical model" would be counted in the larger search.

Although this table represents only a small sample of the literature, it appears that the role of mathematics in the biological sciences has increased. This should not be a surprise, as mathematics gives extra predictive power beyond mere observations, and scientists want this predictive power. Researchers in these three aspects of mathematical biology (such as the book's second author) can vouch for the fact that many articles are written by biologists who use mathematics to explain and extend the power of their results, and many are written by mathematicians inspired to study a biological system. Many are written by interdisciplinary teams.

Google Scholar search term counts.

Search term No specified time period	Count	Mathematical search term	Count	Percent mathematical
"ecology"	3,540,000	"mathematical model ecology"	174,000	4.9
"malaria"	1,680,000	"mathematical model malaria"	55,000	3.3
"tumor"	3,730,000	"mathematical model tumor"	273,000	7.3
Since 2016:				
"ecology"	110,000	"mathematical model ecology"	27,000	24.5
"malaria"	56,100	"mathematical model malaria"	5,270	9.4
"tumor"	144,000	"mathematical model tumor"	17,500	12.2

Not all models are of equal quality of course. Some give predictions a practitioner might well believe, and others are no better than guesses.

Those who wish to read, understand and use the insights gained in this research need to think critically and be knowledgeable consumers of this quantitative information. This book is intended to equip such a person with the tools and intuition to be able to carry out this kind of work.

How we have used this book

Both authors have taught this material to a mixed class of biology students with little math background and also math majors who have taken at least linear algebra and often differential equations. We can verify that both groups come to the course with remarkably little recollection of what they learned in either their biology or math classes. Colleges and universities typically require math courses, pre-medical education usually includes calculus and biology, and more recently quantitative reasoning courses are required at the college level or within a discipline. Such courses will do little good in the long run if the material learned is not remembered or cannot be transferred to new problems. So an important pedagogical problem that this book addresses is this: How can we engage students from both of these groups in a way that fosters critical and creative thinking, and that sticks with the student after the course is over?

Most calculus books and courses help students do the following:

(1) learn some new technique,
(2) apply it to a series of practice problems,
(3) take a test on it, then
(4) go on to the next new technique,

Most biology books and courses help students do the following:

(1) read a huge amount of information,
(2) use some of it in labs,
(3) take a test on it, then
(4) learn more information

We wanted to do design quite a different class and to help both kinds of students have a more meaningful experience. The book (and the courses we have taught from it) shift the emphasis to different pedagogical approaches. So when we have used the book in class, our interest is in turning students into independent researchers and putting them in charge of posing and answering their own research problems.

After a lifetime of being handed math problems to solve, any student might be forgiven for experiencing alarm at the question, "What problem would you like to solve?" And yet forceful arguments have been made that learning that sticks is exactly the learning based on this question. Research experiences for undergraduates are known to improve learning and retention, but usually these are relegated to internships and summer programs. By creating a rich experience for students right in the classroom we hope to give them a chance to ask a question of interest to them, find their own unique answer to it, work with a team of enthusiastic peers, and write a paper of which they can be truly proud. All of these experiences are built to tie the mathematics and the biology to students' own emotions and motivations, thereby causing it to stick.

The roles of classroom actors

Actor	Role in traditional format	Role in research based format
Student	Absorbs assigned information, learns assigned computational techniques and reasoning. Takes tests.	Poses the research question and learns material necessary to solve problem. Writes research papers.
Teacher	Explains and describes, sets tasks to be completed, judges performance on intermediate tasks.	Asks additional questions, critiques thinking, helps modify approach, is a member of every team, does not judge the process of development but only the outcome.
Content delivered in class or textbook Textbook	Information to be tested on. A resource for everything to be learned in the course. The whole mountain.	Ideas that might be helpful in students' own research. A platform from which students begin to form questions and strategies. Base camp.
Assessment	Tasks set by instructor in homework and examinations on course material.	Research papers by groups of students on just about anything to do with biology.
Research literature	Peripheral (to learning)	Central (to research)
Final grade	Examinations made with the express intention that not everyone will get the same grade.	Papers judged against a standard, not against each other.

Students as researchers

In our classrooms the roles played by the various actors differ considerably from traditional educational forms, summarized in the table below.

By concentrating on getting the best possible solution to their own research problem, students naturally encounter plenty of small problems in quantitative reasoning that arise naturally as they try to make the data from any experiment or field study relevant to their mathematical model. A deeper understanding of the meaning and importance of calculus happens equally naturally in the context of building systems of differential equations. They become critical evaluators of the published papers they are using to study their problem. They have an interdisciplinary research experience, without extra cost to them, a funding organization, or an institution. A course taught from this book can give a student with interests in biology a clear and tangible understanding of the value of quantitative reasoning, in general, and calculus, in particular, in the posing and solving of problems.

As an additional outcome, our attitude toward teaching is completely altered by this approach. It our privilege to work with these students. We nearly always learn something new from them. We get an overview of potential research areas we would not have thought about otherwise. We build a base of future colleagues with whom we may write papers in the future. All of these things compensate greatly for the additional time and attention this sort of teaching requires. We hope this book will lighten the burden of teaching this kind of course.

Some details about the book

The book is divided into five parts. The first part is some background on the Lake Victoria region and some review of calculus. The review of calculus focuses on terminology and physical intuition. The next four parts cover four general areas in biology in which biologists and mathematicians alike have posed and solved problems using the techniques of calculus. The four parts consist of three different kinds of chapters. Some chapters explain the biology and mathematics required to understand the kinds of problems posed in the fields covered by the book, some chapters discuss computational techniques used to solve problems, some chapters introduce the reader to what it is like to carry out research and some chapters develop the reader's ability to think like a mathematical modeler. The variety of fields within biology covered by the book as well as the variety of the kinds of things the reader is being asked to read and think about speak to the deeply interdisciplinary nature of the material and our approach to it.

Almost every chapter contains its own bibliography to allow the reader a springboard for finding related work that may be of interest. The book contains three different kinds of tasks (in addition to reading) for reader to engage in. Interspersed throughout the reading are short, contextualized questions to help the reader check their own understanding. If the reader cannot give a somewhat immediate answer to these "Exercises", then we recommend the reader back up and make sure that they have understood everything up to that point in the chapter. At the end of most chapters we have "Problems for Exploration," which, as in most textbooks serve as practice for the kind of tasks we want students to be able to do but, unlike most textbooks, the problems and tasks are practice for the carrying out of research. The questions are often open ended and require the application of the general intellectual and modeling skills we are trying to develop in the reader. Finally, at the end of each part, we present sample projects that the reader could undertake. In our experience, something interesting will result if the reader takes an existing paper in the literature and "tweaks" it in some small way. The projects we present are in that spirit and are provided to show what students in this class have been capable of.

The programming language highlighted in this book is Sage, but the computational methods required to carry out these projects is available in other computer algebra systems and mathematical programming languages. We chose Sage because it is a Python library and, as such, is fairly readable and intuitive. The sample code and all the plots in the book are available for download at

https://github.com/nathancryan/math-bio-book

Conclusion

We acknowledge that this book is unconventional but we believe that this kind of book and the kinds of things it asks of its readers can impact the way its readers connect with mathematics and biology. For readers who are biologists, they will be given tools to understand biological questions in ways that they otherwise would not be able to. For readers who are mathematicians, they will be shown a new landscape in which their way of understanding the world can have an impact. For readers who are neither, we hope that they will appreciate a guided and interactive tour through a truly interdisciplinary field.

All scientists have three main tasks: Problem Posing, Problem Solving and Peer Persuasion. We believe that this book and courses taught in the way that we have described will train students to carry out these three essential tasks.

Acknowledgments

We would like to thank the reviewer at World Scientific Publishing for a careful reading of a draft version of this book and the editors for supporting us throughout the process. We would also like to thank Matt Mizuhara for helpful feedback.

Additionally we would like to thank the National Science Foundation for supporting the Math Across the Curriculum project at Dartmouth College (1994–2001) that led to an early version of this text.

Part 1

Background

CHAPTER 1

Lake Victoria

Lake Victoria (also known as *Nam Lolwe* in Luo; *Nalubaale* in Luganda; *Nyanza* in Kinyarwanda and some Bantu languages) is one of the African Great Lakes. With a surface area of 68800 km^2 [13], it is the largest lake in Africa and the largest tropical lake in the world. Among freshwater lakes it is second largest; Lake Superior in the United States is the largest.

Three countries have shoreline on Lake Victoria: Kenya has about 6% of the shoreline, Uganda about 45% and Tanzania the remaining 49% [11]. The Lake is drained only by the Nile River and is filled by direct rainfall and a large number of streams. As such, its catchment is quite large and includes, in addition to Kenya, Uganda and Tanzania, the nations of Burundi and Rwanda.

In this chapter we give a necessarily *very* brief overview of various aspects of Lake Victoria and its surrounding areas, emphasizing those that are revisited and studied later in this book. There are many useful references for understanding what we describe below. We relied on the informative and thorough [2].

1.1. Humans around Lake Victoria

The Baganda, the native people on the north of the lake, called the lake *Nalubaale* or "Home of the Spirit". The Baganda are a Bantu ethnic group native to Buganda, a sub-national kingdom within Uganda. Traditionally composed of 52 tribes, the Baganda are the largest ethnic group in Uganda, comprising 16.9% of the population. The early history of the people is unclear but, since they are a Bantu-speaking people, it is most likely they originated as a people in Western Africa and moved to their current location during the great Bantu Migration [3].

Long before European exploration of the region, Arab traders were active along the East African coastline and along inland routes in search of gold, ivory, other precious commodities and slaves. An excellent map, known as the *Al Idrisi map*, after the calligrapher who developed it, dated from the 1160s, clearly depicts an accurate representation of Lake Victoria, and attributes it as the source of the Nile. The Arab name for the lake was *Ukerewe*.

The first European to see the lake was the British explorer John Hanning Speke in 1858. Hanning embarked on an expedition with Richard Francis Burton to discover the source of the Nile River. Due to its size, he believed the Lake was the source of the Nile River and so he named it after Queen Victoria. We pause to remark that this name has been a source of frustration to many people in the region and there has long been a general desire to change the name to something less "foreign" [2]. This movement has yet to succeed, in part, because of the diversity of the people who live near and around the Lake and the diversity of names that have been given to the Lake throughout history.

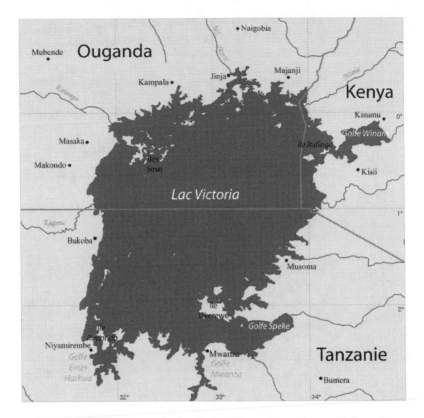

FIGURE 1.1. A map of the Lake Victoria region. Taken from
Wikipedia [5] under reuse license.

The population of the three countries that border the Lake in 2014 was estimated to be 135 million; in 2013 the basin was estimated to have a population of
about 40 million. The largest ethnic groups dwelling around Lake Victoria are the
Luo, Samia, and Suba people in Kenya; the Buganda and Busoga in Uganda; and
the Luo, Kuria, Zanaki, Ruli, Haya, Hangaza, Nyambo, Subi, Sukuma, Kerewe,
Jita, Kara, and Zinza people in Tanzania [2].

Currently the Basin supports one of the densest and poorest rural populations
in the world. The growth rate for the whole Basin is estimated to be 3% per year
and in towns and municipalities the rate is estimated to be 7% per year.

1.2. Challenges facing Lake Victoria

In this book we attempt to model some of the challenges facing Lake Victoria
and the people who live in its basin. The challenges that we will describe mathematically are environmental and medical in nature. In particular, we focus on the
plants and fish that live in the Lake and the diseases faced by the people that live
in the region. We now give some broader context to some of these topics.

Water. The Lake provides a great deal of fresh water to the region. The water
serves many purposes but the ones that are most relevant to us are that it provides
drinking water for both humans and livestock and habitats for various types of flora

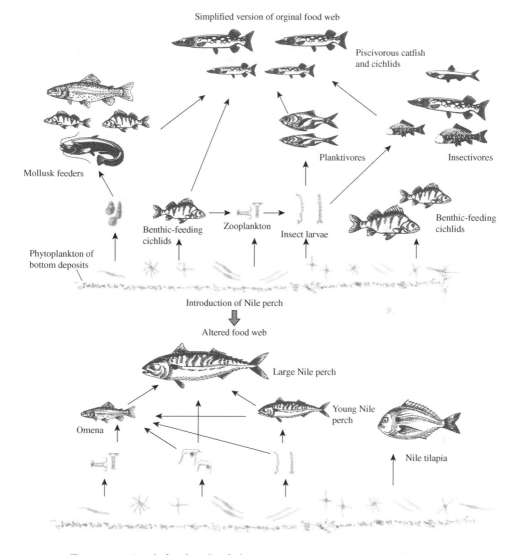

FIGURE 1.2. A food web of the major organisms in Lake Victoria.
The first picture illustrates the food web before the introduction
of the invasive Nile Perch in the 1960s and the second picture
illustrates the food web since the introduction of the fish.

and fauna. In the 1960s the ecosystem of Lake Victoria changed drastically with
the introduction of the Nile perch. See Figure 1.2 for an illustration of the Lake's
food web before and after the introduction of this invasive fish.

The waters of the Lake have been damaged in recent years. Sewage comes
from ever-growing cities, human and animal waste flows into the Lake, and water is
being contaminated as a result of direct uses of the water (e.g., fishing and dams).
The worsening condition of water has caused, among several negative consequences,
increased health risks to humans from tropical diseases such as malaria, cholera,
typhoid, dysentery, and bilharzia.

Fisheries. There is a big demand worldwide for the fish that inhabit the Lake: filleted Nile perch, tilapia and *dagaa* are the most commonly exported. The exportation of fish is a large part of the region's economy. Fishing generates direct income for thousands of people and secondary income for thousands more. In addition, it provides protein-rich food for the people living in the Lake's basin. Fish production in Lake Victoria had reached 400,000 to 500,000 metric tons in 2006, with Tanzania accounting for 40%, Uganda 25% and Kenya 35%.

In all cases, fishing continues largely at an artisanal level with very limited use of outboard engines. In recent years, though, large fishing companies have started appearing in the Lake and there is a fear that the smaller independent artisanal fishermen will be swallowed up by these larger companies. This is relevant to the questions we will ask later because the smaller independent local fisherman have accumulated wisdom and knowledge of the cycles of the fish populations. They know when to fish aggressively and when not. These large companies often have no interest in the conservation of the Lake's fish biodiversity [2, p. 33].

In recent years there has been a protracted decline in the number of fish being caught on the Lake. It is likely that this trend is the result of over-fishing. The Nile perch catches have decreased in number and the individual fishes being caught have decreased in size. More young Nile perch are being removed from the Lake, reducing the population of that fish even more. Additionally, the Nile perch itself has no natural predators and so is proportionally overpopulating the Lake. This leads to them decimating the population of the fish they prey on; now that the Nile perch are being overfished, it is not clear how the populations of smaller fish will respond. The interconnectedness between the fishing industry, the populations of Nile perch and those of smaller fishes is worth studying [2, pp. 37–39].

Health Issues. As mentioned above, the worsening water quality in the Lake due to agricultural runoff, industrial and urban pollution from the Lake's catchment area, has led to a number of health issues. Algal blooms on the Lake lead to a variety of environmental problems (fish kills, more difficult water sanitation, etc.) which can lead directly or indirectly to health problems.

Before the introduction of Nile perch, the Lake was teeming with insectivores. Many diseases in the region are spread by mosquitoes, so anything that reduces the population of insect-eating fish or increases the habitat of mosquitoes leads to negative health effects. An example of the latter is the water hyacinth. This species was introduced to the Lake in the 1980s by human activity and in areas where this weed is most prevalent, there has been a clear increase in the incidence of mosquito-borne illnesses. The plant has been shown to serve as a habitat for several species of mosquitoes [7] and vectors for other diseases. Since the early 2000s there has been a relatively successful management plan in place which makes use of weevils, insects that eat the weed. Additionally, many people in the region have put the plant to good use. For example, it can be used for composting, fodder and silage and biofuel.

In addition to the algal blooms caused by eutrophication, worrisome concentrations of heavy metals (such as lead and mercury) have been found in the Lake, perhaps as a result of industrial activity. See [9] for an overview of heavy metal pollution in Africa. In [10] it is suggested that eating contaminated fish is the main source of lead poisoning for children in Kenya near the Lake. If a child has too high a concentration of lead in their blood they will have depressed development in all areas of development (see, [8], for example) and the effects are irreversible. The medical means to combat lead poisoning in an individual is to lower the concentration of lead in the person's blood through a process called chelation, an expensive procedure not available to poor individuals.

Two other health issues are prevalent in and around the Lake. The first is African sleeping sickness, which is spread via the *tsetse* fly. The fly is very common in and around the Lake, so common that it has rendered many parts of the Lake essentially uninhabitable [12]. The disease has high mortality but can be treated somewhat successfully via a variety of drugs (e.g., pentamidine). See [4] for more information about this disease and how it spreads. The second common health issue is HIV/AIDS. It is estimated that 60% of sub-Saharan Africans under the age of 18 will die from AIDS before they reach the age of 45 [6]. Among people in the Lake region, the incidence of this disease is particularly high among the people who make their living from fishing [1] due, in part, to a general lack of education, how insular the community is and the prevalence of prostitution, formal and otherwise.

1.3. Our small contribution

As seen above, the Lake region and its people are facing a large number of very real and very urgent challenges. In this book we will discuss how to model many of these challenges mathematically. While this is far from anything resembling a means to directly address the challenges faced by the region and its people, we hope that by explaining the required mathematics to future biomedical researchers we might, at some point in the future, make a small contribution to understanding the current situation in the Lake region. At the very least we are raising awareness about an important region in the world. From a scientific perspective, we feel that the rich interplay between environmental and medical issues in this region requires a quantitative way to approach complex systems and is the perfect setting for developing the mathematical tools needed to do so.

In particular, using real data (by and large) collected and published by scientists working in the region, we discuss how to model the following (among others):

- algal growth, water hyacinth growth, and fish growth;
- predator-prey interactions among fish and among insects and plants;
- lead in the bloodstream and the chelation process;
- how diseases such as HIV/AIDS spread through a region.

In this chapter, we hope that we have justified our choice of topics.

Bibliography

[1] Gershim Asiki, Juliet Mpendo, Andrew Abaasa, Collins Agaba, Annet Nanvubya, Leslie Nielsen, Janet Seeley, Pontiano Kaleebu, Heiner Grosskurth, and Anatoli Kamali. HIV and syphilis prevalence and associated risk factors among fishing communities of Lake Victoria, Uganda. *Sexually transmitted infections*, 87(6):511–515, 2011.

[2] Joseph L Awange et al. *Lake Victoria: ecology, resources, environment*. Springer Science & Business Media, 2006.

[3] Philip Briggs and Andrew Roberts. *Uganda*. Bradt Travel Guides, 2010.

[4] John Ford et al. *The role of the trypanosomiases in African ecology. A study of the tsetse fly problem*. Oxford, Clarendon Pr., 1971.

[5] Wikimedia Foundation. Rusing Island. https://en.wikipedia.org/wiki/
     Rusinga_Island, 2015.

[6] Ezekiel Kalipeni et al. *HIV and AIDS in Africa: beyond epidemiology*.
     Blackwell Publishing, 2004.

[7] Noboru Minakawa, Gabriel O Dida, George O Sonye, Kyoko Futami, and
     Sammy M Njenga. Malaria vectors in Lake Victoria and adjacent habitats
     in western Kenya. *PLoS One*, 7(3):e32725, 2012.

[8] Herbert L Needleman, Alan Schell, David Bellinger, Alan Leviton, and Eli-
     zabeth N Allred. The long-term effects of exposure to low doses of lead in
     childhood: an 11-year follow-up report. *New England Journal of Medicine*,
     322(2):83–88, 1990.

[9] Jerome O Nriagu. Toxic metal pollution in Africa. *Science of the Total Envi-
     ronment*, 121:1–37, 1992.

[10] Elijah Oyoo-Okoth, Wim Admiraal, Odipo Osano, Veronica Ngure, Michiel HS
     Kraak, and Elijah S Omutange. Monitoring exposure to heavy metals among
     children in lake victoria, kenya: Environmental and fish matrix. *Ecotoxicology
     and Environmental Safety*, 73(7):1797–1803, 2010.

[11] J Prado, RJ Beare, J Siwo Mbuga, and LE Oluka. *A catalogue of fishing
     methods and gear used in Lake Victoria*. Food & Agriculture Org., 1991.

[12] August Stich. Human African Trypanosomiasis: The Smoldering Scourge of
     Africa. In *Zoonoses-Infections Affecting Humans and Animals*, pages 785–799.
     Springer, 2015.

[13] J-P Vanden Bossche and Garry M Bernacsek. *Source book for the inland fishery
     resources of Africa*, volume 1. Food & Agriculture Org., 1990.

CHAPTER 2

What is Calculus?

Calculus is without a doubt one of the greatest human inventions. It arose from a need for scientists to explain the world and is still used for the same purpose. The text that follows assumes a pretty high level of comfort with the ideas of differential calculus. In particular, the reader will be expected to understand rates of change, the derivative both at a point and as a function, and antiderivatives. Most of this book is about differential equations (which we define in Chapter 3) but, to really understand those, we need to make sure that the reader understands the basic notions of a typical first semester calculus class. We refer the reader to any number of standard calculus books if they feel the need to have a more serious refresher (e.g., [3, 5, 8, 9]).

Since this book is about the intersection of mathematics and biomedicine, we motivate the notions in this chapter via a model of solid tumor growth. See [6, 7] for further information on the dynamics of tumor growth.

2.1. Biological context

We start by giving a very cursory introduction to the biology of solid tumors. Non-cancerous cells grow on their own, with the homeostatic regulatory system (HRS) responsible for maintaining a balance between the creation of cells and the destruction of cells. When the HRS becomes compromised we get cancer. In particular, we get an abnormally accelerated multiplication of cells that might eventually become a tumor.

How can one model the growth of a solid tumor? We now present one hypothetical model, called the Gompertz model. A solid tumor increases in size through the process of angiogenesis: as cell populations increase rapidly, surrounding blood vessels are pulled into the rapidly growing tissue and begin to form new blood vessels. Now that these cells have their own blood supply, they experience unfettered growth for a while. At some point, though, the tumor is hypothesized to outstrip its blood supply, leading to a decrease in growth rate. Mathematically, it will reach some maximal size at which point its growth rate will be effectively zero. Using parameters found in the literature [7] we created a graph of a tumor's growth rate versus the size of the tumor; see Figure 2.1.

2.2. Rates of change

When we measure how things change, we often calculate the *rate of change*. We start by considering the case of tumor growth as we did above. In Figure 2.1, the vertical axis determines the growth rate of tumor cells. How would such data be calculated? Let $N(t)$ be the number of tumor cells at time t, measured in number of cells and days, respectively. What an experimental biologist studying tumor growth

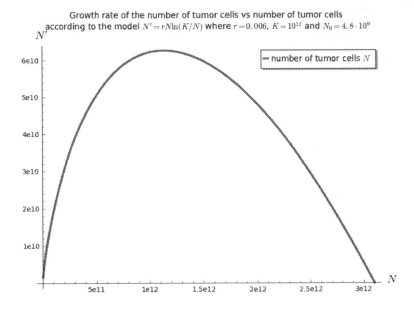

Growth rate of the number of tumor cells vs number of tumor cells according to the model $N' = rN\ln(K/N)$ where $r = 0.006$, $K = 10^{12}$ and $N_0 = 4.8 \cdot 10^9$

FIGURE 2.1. A sample graph of tumor growth rate versus tumor size assuming the hypothetical Gompertz model described in the text. We make the following observations. Until almost the very end, the growth rate is positive and so the tumor is always growing. We also observe that the initial growth rate is almost linear: this suggests exponential growth at the beginning. Finally, we observe that at some point the growth starts decreasing: this corresponds to the point at which the tumor is starting to overtax its supply of nutrients.

in a Petri dish might do is the following: at time $t = 100$ days, the scientist would count the cells and then, maybe 10 days later, the scientist would count cells again. Suppose that $N(100) = 1.62 \times 10^{12}$ cells and $N(110) = 1.68 \times 10^{12}$ cells. Then, the rate of change at 100 days would be approximated by the slope of the secant line connecting the points $(100, 1.62 \times 10^{12})$ and $(110, 1.68 \times 10^{12})$. In particular, we would get a rate of change of

$$\text{ROC} = \frac{\Delta N}{\Delta t} = \frac{\text{final} - \text{initial}}{\text{change in } t} = \frac{1.68 \times 10^{12} - 1.62 \times 10^{12}}{110 - 100} = 0.06 \times 10^{11} \text{ cells per day.}$$

Since the ROC is positive, we observe that the number of cells is increasing after 100 days. On a graph like the one in Figure 2.1, this would correspond to a point $(1.6 \times 10^{12}, 0.06 \times 10^{12})$.

There are several possible reasons for which this number could be wrong. We will not focus on the errors introduced by the skill-level of the individual experimentalist nor on the error introduced by the fact that the counting method cannot possibly be exact. We will focus on the error introduced by the mathematical approximation.

We note that time is continuous–for instance, between any two given times, there is always another time between them. Imagine (we really mean *pretend*) that the population of tumor cells is also continuous. Then, ideally we would like to be able to do the following. We would get a better approximation to the rate we calculated above if we could calculate the number of cells at time 100 and at time 100.1. This would give a rate of change of

$$ROC = \frac{\Delta N}{\Delta T} = \frac{N(100.1) - N(100)}{100.1 - 100}.$$

This would be more accurate than the approximation from before because less variation can happen between 100 days and 100.1 days. Can we do even better? Say, we could compute the rate of change based on $t = 100$ days and $t = 100.00001$ days. That would be even better, right? What's the best we can do? The best would be to calculate the derivative at time $t = 100$ days; i.e., $N'(100)$, also written $\frac{dN}{dt}|_{t=100}$. We introduce this idea in the next section.

2.3. The derivative at a point

We want to get as good an approximation for the rate of change of the population of tumor cells at a certain time t_0 as possible. In other words, we want to compute

$$ROC = \frac{N(t_0 + \Delta t) - N(t_0)}{t_0 + \Delta t - t_0} = \frac{N(t_0 + \Delta t) - N(t_0)}{\Delta t}$$

where Δt, a fancy way of writing "a certain amount of time in the future (or in the past)", is getting smaller and smaller.

The right way to do this, then, is via the limit. In particular, we write

$$\text{Instantaneous ROC at } t_0 = N'(t_0) = \frac{dN}{dt}\bigg|_{t=t_0} = \lim_{\Delta t \to 0} \frac{N(t_0 + \Delta t) - N(t_0)}{\Delta t}$$

where the $\lim_{\Delta t \to 0}$ means that we plug in a sequence of infinitely many Δt just greater than zero and just less than zero. We denote this number by $N'(t_0)$, if it exists (for us it will always exist because we are assuming that our population function is *differentiable*). Of course, we cannot plug infinitely many of anything, but this is the intuitive idea. Another complementary intuitive idea is that we want to plug in 0 for Δt, resulting in a ratio of $\frac{0}{0}$. To give an actual definition of instantaneous ROC, we need to do things analytically.

EXAMPLE 2.1. Let the number of bacteria in a petri dish at time t hours be given by $N(t) = t^2$ (in Chapter 3, you will see the model that actually describes such a population). What is the instantaneous rate of change after two hours? Notationally, we are being asked to find $N'(2)$. Using the formulas from above, we have

$$N'(2) = \lim_{\Delta t \to 0} \frac{N(2 + \Delta t) - N(2)}{\Delta t}$$
$$= \lim_{\Delta t \to 0} \frac{(2 + \Delta t)^2 - 2^2}{\Delta t}$$
$$= \lim_{\Delta t \to 0} \frac{4 + 4\Delta t + (\Delta t)^2 - 4}{\Delta t}$$
$$= \lim_{\Delta t \to 0} \frac{4\Delta t + (\Delta t)^2}{\Delta t}.$$

We are not yet done. Before we finish, we take a small algebra detour. Is it always true that $\frac{x}{x} = 1$? No. For all $x \neq 0$, this is true, but not for $x = 0$. Fortunately for us, when evaluating the limit above, the only value of Δt we are forbidden to plug in is 0: the notation $\Delta t \to 0$ means that Δt gets arbitrarily close to 0 from both the positive and negative sides but never actually takes on a value of 0. So, we can cancel $\frac{\Delta t}{\Delta t}$ should we run into such an expression. Now, continuing with the above calculation,

$$
\begin{aligned}
N'(2) &= \lim_{\Delta t \to 0} \frac{4\Delta t + (\Delta t)^2}{\Delta t} \\
&= \lim_{\Delta t \to 0} \left(4\Delta t + (\Delta t)^2\right) \cdot \frac{1}{\Delta t} \\
&= \lim_{\Delta t \to 0} \frac{4\Delta t}{\Delta t} + \frac{(\Delta t)^2}{\Delta t} \\
&= \lim_{\Delta t \to 0} 4 + \Delta t \\
&= 4,
\end{aligned}
$$

where the last limit follows from the intuitive fact that 4 plus something arbitrarily small is 4. The answer, in the context dictated by the problem, should be interpreted as "at 2 hours, the population of bacteria is growing at 4 bacteria per hour."

To really understand analytic calculations like the ones above, the reader should really have a firm understanding of the definition of a limit and theorems related to the computation of limits. We refer the reader without such a background to one of the calculus books in this chapter's bibliography.

The value $N'(t_0)$ is the slope of a line. Namely, it is the slope of the line tangent to the function $N(t)$ at time t_0. In other words, if you were placing a ruler along the graph of $N(t)$ at $t = t_0$, in such a way that the ruler was going roughly in the same direction as the graph at $t = t_0$ and only touch the graph at one point, the slope of the ruler would be $N'(t_0)$.

2.4. The derivative as a function

In the above we discussed how to compute $N'(t_0)$, the rate of change of a population $N(t)$ at a particular fixed time t_0. It would be nice to not have to repeat the whole process we described for each and every time we might care about. Instead we think of $N'(t)$ itself as a function of t and carry out one calculation for an arbitrary time t. The result will be a function of time t and given a particular value of t we care about, say time t_0, all we would need to do is to plug t_0 into our function $N'(t)$.

EXAMPLE 2.2. Let the number of bacteria in a petri dish at time t hours be given by $N(t) = t^2$ (again, in Chapter 3, you will see the right way to model this). What is the instantaneous rate of change after t hours? Notationally, we are being asked to find $N'(t)$, for an arbitrary time t. Using the formulas from above, we have

$$
\begin{aligned}
N'(t) &= \lim_{\Delta t \to 0} \frac{N(t + \Delta t) - N(t)}{\Delta t} \\
&= \lim_{\Delta t \to 0} \frac{(t + \Delta t)^2 - t^2}{\Delta t} \\
&= \lim_{\Delta t \to 0} \frac{t + 2t\Delta t + (\Delta t)^2 - t^2}{\Delta t}
\end{aligned}
$$

$$= \lim_{\Delta t \to 0} \frac{2t\Delta t + (\Delta t)^2}{\Delta t}$$

$$= \lim_{\Delta t \to 0} \left(2t\Delta t + (\Delta t)^2\right) \cdot \frac{1}{\Delta t}$$

$$= \lim_{\Delta t \to 0} \frac{2t\Delta t}{\Delta t} + \frac{(\Delta t)^2}{\Delta t}$$

$$= \lim_{\Delta t \to 0} 2t + \Delta t$$

$$= 2t,$$

where the last limit follows from the intuitive fact that $2t$ plus something arbitrarily small is $2t$. The answer, in the context dictated by the problem, should be interpreted as "at t hours, the population of bacteria is growing at $2t$ bacteria per hour."

We hope that the reader has observed that this calculation is essentially identical to the one in the previous example with the only difference being that instead of carrying a 4 through the calculation, we carry a $2t$. We do this because we find that students get nervous when they are asked to think of the derivative as a function. The transition from the derivative at point to the derivative as a function is somehow conceptually more difficult. We think that repeating essentially the same calculation in both cases will help convince the reader there is no reason to be worried.

Finally, we hope the reader can observe that the results of these two examples are compatible: namely $N'(2) = 2 \cdot 2 = 4$.

Thinking of the derivative as a function has many advantages, not the least of which is that it saves us the work of separately calculating $N'(t_0)$ for every t_0 we need to know about. Another advantage is if we know $N'(t)$ is some function and we know something about this function, we can conclude something about the growth of $N(t)$.

EXAMPLE 2.3. In the above we calculated $N'(t)$ for $N(t) = t^2$ and determined $N'(t) = 2t$. How is this useful? Observe $N'(t) > 0$ when $t > 0$. What does this mean? The derivative being positive means that the rate of change is positive which, in turn, means that the function $N(t)$ is increasing when $t > 0$. This is consistent with the graph of $N(t)$ (we trust the reader either has a clear idea of what the function $N(t) = t^2$ looks like or will take the time to graph the function for themselves). Similarly, when $t < 0$ we have $N'(t) < 0$, or that $N(t)$ is decreasing. Finally, when $t = 0$, we have that $N'(t) = 0$. This means that the tangent line at $t = 0$ would be horizontal, meaning that the function is neither increasing nor decreasing. Again, this is consistent with graph of $N(t)$.

See Table 2.1 for a table of common functions and their derivatives.

2.5. Antiderivatives

In mathematics we often want to be able to undo the result of an operation. For example, raising a number to the rth power and taking the rth root. The same holds for derivatives. It will sometimes be useful to be able to do the following. Given a function $f(t)$ which we know to be the derivative of some other function $F(t)$, can we find $F(t)$?

$f(t)$	$f'(t)$	$f(t)$	$f'(t)$
t^r	rt^{r-1}	$\sin t$	$\cos t$
e^t	e^t	$\cos t$	$-\sin t$
$\ln t$	$\frac{1}{t}$	$cf(t)$	$cf'(t)$

TABLE 2.1. A table of common functions and their derivatives. We recall, too, that the derivative of a sum is the sum of the derivatives and that the analogous fact does not hold for products and quotients. We observe that the r in the first entry can be any real number.

EXAMPLE 2.4. Suppose we are given $f(t) = (t+2)^2$ and we are asked to find the $F(t)$ so that $F'(t) = f(t)$. The first observation we make is that there is more than one correct answer. Since the derivative of any constant is zero (see the first entry of Table 2.1 with $r = 0$) we see that if $F'(t) = f(t)$ so is $G'(t)$ where $G(t) = F(t)+c$ for any constant c. Now we carry out the computation (without, as some readers might expect, using the Chain Rule). Note that

$$f(t) = (t + 2)^2 = (t + 2)(t + 2) = t^2 + 4t + 4.$$

We want an $F(t)$ so that $F'(t) = t^2 + 4t + 4$. We proceed term by term. The derivative of what gives us a t^2? Looking at the rules in Table 2.1 we see that the derivative of a power function x^r reduces the exponent by one. So our first guess for the antiderivative of t^2 is t^3 but we see that this cannot be correct. Since the derivative of t^3 is $3t^2$, we need to adjust our first guess to handle the extra three that pops up. So, if our first guess is t^3, we make our second guess for the first term to be $\frac{1}{3}t^3$. We see that the derivative of this is indeed t^2, as desired. Continuing this way, we see that *an* antiderivative of $f(t)$ would be $F(t) = \frac{1}{3}t^3 + 2t^2 + 4t + 1000000$. It is easy to check if we have this right. By taking the derivative of $F(t)$ we see that $F'(t) = f(t)$ and so we have found an antiderivative of $f(t)$.

The $\int$ symbol indicates that we want to find an antiderivative. For instance, in the example we just considered, we could have written

$$\int (t + 2)^2 \, dt$$

where the dt tells us that t is the variable of the function (so that we do not confuse it with any constants floating around). It turns out that every antiderivative of a given function $f(t)$ differs from another by a constant. So, it is customary to write

$$\int f(t) \, dt = F(t) + C$$

where $F'(t) = f(t)$ and C is a constant.

We can also define the (definite) integral $\int_a^b f(t) \, dt$ which is the signed area of $f(t)$ between $t = a$ and $t = b$. It turns out that this signed area, by the Fundamental Theorem of Calculus, can be calculated by evaluating the antiderivative of $f(t)$ at the end points $t = a$ and $t = b$. (Of course this is provided the antiderivative exists which it will in this book since every function $f(t)$ we consider will be assumed to be continuous.) In particular,

$$\int_a^b f(t)\ dt = F(b) - F(a)$$

where F is any function such that $F' = f$.

2.6. Fitting a tumor growth model to data

Now, having reviewed the calculus that we will need in this book, we return to the tumor growth model. Recall Figure 2.1. With the notation we have in hand now, this curve is the graph of $N'(t)$ versus $N(t)$ given by the following rule:

$$(2.1) \qquad\qquad N'(t) = rN(t) \ln\left(\frac{K}{N(t)}\right).$$

This is the first of many differential equations in the book. A differential equation is, roughly speaking, an equation relation functions that their derivatives and whose solution is a function that makes the equation true. In this case, we would be looking for a function $N(t)$ whose derivative is $rN(t) \ln\left(\frac{K}{N(t)}\right)$. It turns out that there will be an infinite number of such functions parametrized by a single parameter.

A task for researchers is to figure out values for K and r. In [2] K was taken to be 10^{12}. In a paper by Norton [4], it was assumed that $N(0) = 4.8 \times 10^9$, $r = 0.055$ and a lethal tumor level of 3.1×10^{12}. Using these assumptions and applying them to survival curves of patients [1], Norton got a perfect visual match between his modeled survival curves and that data. The assumption that $N(0) = 4.8 \times 10^9$ helps us identify which of the infinite number of solutions we want to pick by helping specify the value of the parameter mentioned above.

Often (2.1) is rewritten, using basic log rules, as

$$\frac{dN}{dt} = rN(\ln K - \ln N) = N(r \ln K - r \ln N) = N(\ln(K^r) - r \ln N) = N(b - a \ln N)$$

where $b = \ln(K^r)$ and $a = r$ are constants. Now, we can ask some questions about the function N and its derivative $\frac{dN}{dt}$.

First, we try to figure out when $N(t)$ is increasing and when it is decreasing. We start by finding when $N'(t)$ is 0. We observe that $N > 0$ (since you cannot have negative tumor cells or zero cells in a tumor) and conclude that $N'(t) = 0$ when $(b - a \ln N) = 0$. This happens when $N = e^{b/a}$, using basic log rules. We call this N_e because it is the population at which the tumor cell population is at equilibrium: since at this population $N'(t) = 0$, the population is neither growing nor shrinking. For our particular example, this will happen as follows:

$$r = 0.006,\ K = 10^{13} \Rightarrow a = 0.006,\ b = \ln(K^r) = 1.80$$

$$\Rightarrow N_e = e^{b/a} = 10^{13}(= K),$$

agreeing with the graph in Figure 2.1.

Second, we try to sketch $N(t)$ based on our knowledge of its derivative. We start by observing (see the graph in Figure 2.1) that up to the population N_e the derivative $N'(t)$ is always positive and so $N(t)$ is always increasing. We then observe that from the beginning of the graph up to around $N = 3.68 \times 10^{12}$, $N'(t)$ is increasing. That is, the population growth is not just getting faster but that is also accelerating. Then, after the population reaches 3.68×10^{12} the rate of increase $N'(t)$ starts decreasing. Finally, we observe that as t gets bigger and bigger, the population should stabilize at around $N_e = 10^{13}$ cells. It is also possible to solve for N analytically; the solution is given in Figure 2.2.

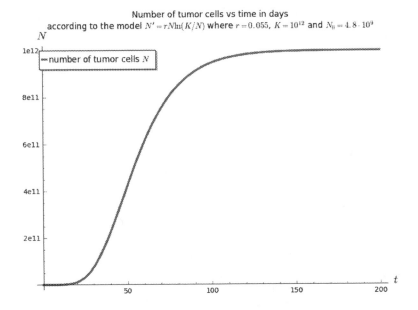

FIGURE 2.2. The graph of $N(t)$ with parameter values $K = 10^{12}$ and $r = 0.0550$. We also assume an initial population of 4.8×10^9 cells. The general shape of this graph can be predicted by the equation $N'(t) = rN \ln \left(\frac{K}{N} \right)$.

2.7. Conclusion

The point of this chapter was to review the basics of differential calculus. Writing such a chapter is a challenge–include too little and it might not be useful to many readers and include too much and it would be a whole calculus book unto itself. We hope that the chapter will have helped re-familiarize the reader with the basic ideas of differential calculus. In particular, the reader should know what a rate of change is, what the derivative is at a point and as a function, and how to read information about a function from the graph of its derivative.

Although some models can be solved explicitly, such as this one, most of the models we will consider in this text are too complex for explicit solutions and must be approached with numerical (computer assisted) tools. Finding explicit solutions to calculus problems is therefore not emphasized. Even so, understanding the basic relationship between a function and its rate of change is a necessary part of setting up and critically examining any model.

Finally, we point the reader to standard introductory calculus books should this chapter not have proved useful to them.

Bibliography

[1] Harris Julian Gaster Bloom, WW Richardson, and EJ Harries. Natural history of untreated breast cancer (1805-1933). *British Medical Journal*, 2(5299):213, 1962.

[2] Roberto Chignola and Roberto Israel Foroni. Estimating the growth kinetics of experimental tumors from as few as two determinations of tumor size: implications for clinical oncology. *IEEE Transactions on Biomedical Engineering*, 52 (5):808–815, 2005.

[3] Deborah Hughes-Hallett, Patti F Lock, Andrew Gleason, Daniel E Flath, and Jeff Tecosky-Feldman. *Applied Calculus 5th ed.* John Wiley & Sons, 2014.

[4] Larry Norton. A gompertzian model of human breast cancer growth. *Cancer Research*, 48(24 Part 1):7067–7071, 1988.

[5] Jon Rogawski. *Calculus*. Macmillan, 2011.

[6] Linda Simpson-Herren and Harris H Lloyd. Kinetic parameters and growth curves for experimental tumor systems. *Cancer Chemotherapy Reports. Part 1*, 54(3):143, 1970.

[7] Sabrina L Spencer, Matthew J Berryman, Jose A Garcia, and Derek Abbott. An ordinary differential equation model for the multistep transformation to cancer. *Journal of Theoretical Biology*, 231(4):515–524, 2004.

[8] James Stewart. *Calculus: early transcendentals*. Cengage Learning, 2015.

[9] Silvanus P Thompson and Martin Gardner. *Calculus made easy*. St. Martin's Press, 2014.

Part 2

Population modeling

CHAPTER 3

Introduction to Population Modeling

This chapter marks the real beginning of the book. In it, we start our discussion of population modeling, paying particular attention to the population of algae in Lake Victoria. As we saw in Chapter 1, and will now see in more detail, algae populations have played several roles in the ecological history of Lake Victoria and so it seems a good place to start.

When ecologists study an ecosystem as complex as the one in Lake Victoria, they often try to write mathematical models to attain one of the two complementary goals:

(1) **Understand how the system changes with time.** The models an ecologist develops are used to predict future changes in this or a similar ecosystem or, alternatively, to hypothesize about the conditions of the ecosystem in the past.

(2) **Understand the important factors in the system.** After studying an ecosystem for some time, an ecologist may be able to predict which factors are most important to the specific system. The ecologist will incorporate data about these factors into the model and then could compare the output to empirical data about the system as a whole.

(3) **To test hypotheses about the system.** The first two items might be carried out with the testing of a particular hypothesis in mind.

While these are the ultimate goals, at least at first, an ecologist may make certain assumptions about the system and its interactions in an attempt to make it more manageable to model. For instance, one may assume that the system is closed (there is no gain or loss of resources or biomass). In some systems this is a reasonable (but not completely accurate) assumption at least for a small amount of time. Even if a system is not closed, there may be a net flow of a resource that is small enough that it can be safely ignored; i.e., the amount gained may approximately equal the amount lost and the difference is negligible.

An ecologist may try to represent pieces of the ecosystem separately and then later try to incorporate some of these many interactions into a more general model. In doing so, one is able to model carefully how a few organisms interact and gain some understanding of how different elements of the system work before the model gets too complex to be usefully analyzed. The ecologist may try to simplify some of the models by making assumptions that certain conditions are true (at least during a reasonable interval of some other factor such as time), or try looking at a group of organisms as one organism (especially if they all belong to the same trophic level) to see what the overall system will tend to do.

3.1. Biological context

In this section we start with a model of algae in Lake Victoria. Algal concentrations are three to five times higher now than during the 1960s. Much of the lake bottom, due to the lake surface being so densely covered by algae, experiences prolonged periods of anoxia, a state in which there is an absence of oxygen reaching the organisms living at the bottom of the lake.

The Lake has suffered from eutrophication, an over-abundance of nutrients. The causes of this include:

- enhanced effluent discharge
- runoff and storm water and
- enhanced discharge of solids.

Eutrophication has led to a variety of other issues in addition to the algal blooms we discuss here (see, e.g., [8] for more details).

In the Kenyan waters of the Lake, algal blooms are widespread and occur throughout the year. The most intense blooms occur between April and July, during the rainy season, and between September and November, during the dry season. The majority of algae blooming in the Nyanza Gulf, an extension of the Lake into Kenya, are blue-green algae, the *Cyanophyceae* group and *Microcystis aeruginosa*, *Anabaena* spp. and *Lyngbya* spp. In the open waters of the Kenyan sector of Lake Victoria, the major blooming algae are the diatoms *Nitzschia acicularlis* [3].

Because of these blooms, the bottom sediments are a thick layer of green mud derived from the decomposing algal blooms. Near this muddy bottom, at depths greater than 25 meters, the water is deoxygenated to the lowest level that can be tolerated by the majority of fish species that live in the Lake. This pushes the fish to a well-oxygenated layer of water near the surface and thus drastically reduces the size of the habitat suitable for fish survival.

In addition to a greatly reduced habitat for fish, the algal blooms are also correlated with massive fish kills [9]. They also indirectly bring about public health issues by taxing and damaging the water treatment facilities in cities near the Lake due to the clogging of filters. Finally, they wreak economic havoc by making the water unsuitable for consumption by livestock and humans and unsuitable for recreational use.

Understanding the factors that control the growth of these algae, then, is a natural first question to ask about the Lake Victoria ecosystem.

3.2. The model

If we want to model the observed algae bloom in terms of growth of the population of algae in Lake Victoria, as described above, we might start by making some simplifying assumptions:

- Resources (such as essential nutrients) are unlimited. This is a reasonably valid assumption at least for a small amount of time when the amount of available resource is large relative to the population of algae.
- There is no migration of algae into or out of Lake Victoria.
- The rates of reproduction and death (natural and due to predation) are constant.

To begin the process of developing a model, it is useful to fix some notation and units. Let N represent the population of algae. A first question might be "How does one measure the population of algae?" Instead of counting or estimating the number of plants, scientists measure the concentration of chlorophyll in the water. If the concentration is above 40 $\mu g/L$, the body of water is said to be in the midst of an algal bloom and concentrations of greater than 100 $\mu g/L$ are associated with fish kills [5]. So, we think of N as the population of algae, but in reality it is the concentration of chlorophyll in the water.

Now, as we saw in Chapter 2, the change in the population of algae is denoted ΔN and we care about the rate of change of the population of algae during a certain period of time that we denote Δt, measured in years. In other words, we want to understand
$$\frac{\Delta N}{\Delta t} \text{ with units } \mu g/L \text{ per year.}$$
The net rate of change of the population of algae is the difference between the rate at which algae are being added to the lake by reproduction and the rate at which it is being removed from the population by death. In other words,
$$\frac{\Delta N}{\Delta t} = (\text{ rate of algae being born })-(\text{ rate of algae dying}).$$

A large number of algae means that more algae can be born, and so the rate at which algae is being born is proportional to the population (N) of algae. Thus, this quantity can be written as
$$(\text{ rate of algae being born}) = bN$$
where the constant b has units of year^{-1} in order for the units to work out. Similarly,
$$(\text{ rate of algae dying}) = dN$$
where the constant d also has units year^{-1}; we give an conceptual explanation for this below, but for now we can see that the units of b and d have to be year^{-1} in order for the units to agree. Then, first by substitution and second by the distributive law we get,
$$\frac{\Delta N}{\Delta t} = bN - dN = (b-d)N.$$

Now we pause a second here to understand these constants and their units. First, what happens if $b = d$? Then the change in population over time is zero and so the population is said to be in equilibrium. This is not likely to be the case, so we ask what if $b > d$? This means that the rate of change of N is positive (because N itself, being a concentration/proxy for algae population, is positive) and thus the population of algae is growing. Algebraically this also makes sense because the constant b should be thought of as being the birth rate and d should be thought of as being the death rate and if the birth rate is greater than the death rate, then the population should be growing. A similar analysis can be done with the case when $b < d$.

Now, we explain the units of b and d–this weird-looking year^{-1}. These constants can be thought of as follows. First, we consider b: this number is what fraction of an algae plant reproduces per unit time, in this case, per year. Second, we consider d: this number is what fraction of an algae plant dies per year. (Again, this should be stated in terms of chlorophyll concentration but we find it easier to talk and think about the algae reproducing instead of the chlorophyll increasing.) Since the algae population has increased over the last 50 years in Lake Victoria we know $b - d$ is positive and can think of this constant as measuring how much faster algae is

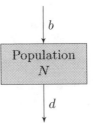

FIGURE 3.1. A box model for a population with exponential growth. This represents a population with unlimited resources.

reproducing than it is dying off, per year. It is the *per capita* growth rate, as can
be seen from this rewriting of the above:

$$\frac{\text{algae}}{\text{year}} \frac{1}{\text{year}} = \frac{\Delta N}{\Delta t} \frac{1}{N} = b - d.$$

The units of $b - d$ can really be thought of as being algae per year, per algae.
Overall, these units simplify to being year^{-1}.

Now, since $(b - d)$ is positive it can be replaced by a positive constant, r, which
will represent the overall rate of population increase per algae in the lake:

$$\frac{\Delta N}{\Delta t} = rN.$$

If we want to determine how the population of algae is changing at a particular
time (determine its instantaneous rate of change), we would proceed as in Chapter 2.
In particular, we would take the limit of $\frac{\Delta N}{\Delta t}$ as the period of time gets infinitesimally
small to obtain the derivative of N with respect to time, resulting in the differential
equation:

$$\frac{dN}{dt} = rN.$$

In mathematical terms this equation states that the rate of change of the population
at a given time is directly proportional to the population at that time.

We can visually represent this equation as in Figure 3.1. The system has a
total population of N algae and this population feeds back into the population at
a rate of r. While this box diagram is particularly simple, we will return many
times to this way of thinking as the models get more complicated (e.g., as they get
more compartments, as they have different rates in each direction between multiple
compartments, etc.).

In population studies, a differential equation usually describes relationships
between a function and its derivatives, and gives us a way of mathematically representing what rules we believe govern how a function such as population changes
over time. Sometimes the equations enable us to see trends in a function even if we
are unable to determine the function itself.

In the situation we are studying, $\frac{dN}{dt}$ represents the rate of change of the population of algae (N). If $\frac{dN}{dt}$ is positive, the rate of change of the population is
positive and therefore the algae population is growing. Similarly, if $\frac{dN}{dt}$ is negative,

the size of the population of algae is decreasing. The magnitude of $\frac{dN}{dt}$ tells us how quickly the size of the population is changing.

If we can figure out how the population, N, changes with time, then we can determine the size of the algae population at any time, t. We want to solve the differential equation

$$\frac{dN}{dt} = rN.$$

by find finding the function $N(t)$ that makes the equation true.

There are several strategies we can use for approaching a model of this sort. One is to attempt to find what scientists call an "analytic" solution to the problem. This means finding all the functions, $N(t)$ which solve the differential equation describing our model. Such a solution is very satisfying and allows one to analyze both the implications and the reliability of the model quite efficiently. Most biological models are too complicated to solve analytically, but the model we are using for algae growth is so simple that an analytic solution is very easy to calculate.

Using a technique called separation of variables we rearrange the terms of the equation:

$$\frac{dN}{N} = rdt.$$

We then take antiderivatives:

$$\int \frac{dN}{N} = \int rdt$$

to obtain

$$\log|N| = rt + c$$

(in this book log always means natural log, i.e., the logarithm in base e). Since N is a population size and is therefore always non-negative, we may remove the absolute value:

$$\log N = rt + c.$$

Next, we solve for N:

$$N = e^{rt+c}.$$

The size of the algae population varies depending on when it is observed, and can be written in such a way to make the dependence of N on t explicit:

$$N(t) = e^{rt+c}.$$

Simplifying a little by using the laws of exponents, we deduce:

$$N(t) = e^{rt+c}$$
$$= e^{rt}e^{c}.$$

At the initial time $t = t_0 = 0$ (we set this time to be zero arbitrarily, we think of zero as the first time that we actually cared about the algae population), the population of algae is some value, N_0 (again, technically this is a concentration). So,

$$N(0) = N_0 = e^{r \cdot 0}e^{c} = e^{c}.$$

Putting everything together we get,

$$N(t) = e^{c}e^{rt} = N_0 e^{rt}.$$

Again, we use units to provide a consistency check for what we have done so far. Recall, the units of r are years^{-1}. Observe that since the units of $N(t)$ and

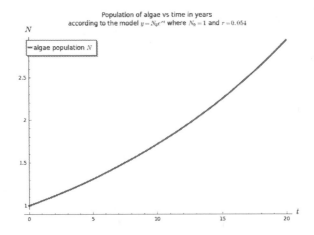

FIGURE 3.2. We see the typical J-shape of exponential (Malthusian) growth in the graph of $N(t) = N_0 e^{.054t}$ with $N_0 = 1$.

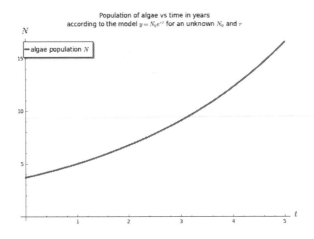

FIGURE 3.3. Figure to be used in Exercise 3.1.

N_0 are the same (they measure the size of the algae population or, technically, a proxy for it), the exponential e^{rt} must be unitless and, in particular, the exponent rt must be unitless, too. Since the units of t are years, we see that the units of r must be years^{-1}.

An equation in this form is commonly known as the *Malthusian exponential growth model*, named after the man who first proposed it as a model for human population growth at the end of the 18th century. If there are sufficient nutrients in Lake Victoria (because of pollution from industry and agriculture) we would predict that this model would give us a fairly good approximation of what is really happening to the algae. If surplus nutrients continue to be added to the lake and if this model is an accurate model of the algae in Lake Victoria, we should expect a continued exponential increase in the number of algae over time.

EXERCISE 3.1. Suppose we gave you the graph in Figure 3.3. Can you determine (or at least estimate) the initial population and r, the rate of growth? Be sure to give units.

According to some sources (e.g., [7]), between 1950 and 1980 the algal biomass production (as determined by chlorophyll concentration) had increased somewhere from five- to ten-fold in the open surface waters. Using our model, if there is a five-fold increase in algae over 30 years, we represent the initial amount of algae as N_0 and the five-fold increase as $5N_0$:

$$5N_0 = N_0 e^{30r}$$
$$\Rightarrow 5 = e^{30r}$$
$$\Rightarrow r = 0.054.$$

Thus, if our simplifying assumptions are valid and the algal biomass has grown in the way described above, the equation

$$N(t) = N_0 e^{0.054t}$$

will determine how much algae there is at time t. A graph of this equation if we choose $N_0 = 1$ is shown in Figure 3.2.

EXERCISE 3.2. Using the growth rate we just found, describe an experiment to calculate (or at least estimate) N_0, the initial population of algae in Lake Victoria.

EXERCISE 3.3. If a population doubles every 5 hours, what is its rate of increase? Be sure to give units.

EXERCISE 3.4. If a population triples every 5 hours, what is its rate of increase? Be sure to give units.

EXERCISE 3.5. Suppose that the rate of increase for a particular bacterium is $r = 0.24$ days^{-1}, how long does the bacterium take to double its population?

3.3. Implications

If the algae population does actually increase by nearly 50 times its original size in the near future, there may be many problems in store for Lake Victoria. Algae blooms not only affect the transparency of the lake water, but they have already greatly increased the mass of decaying matter and resulted in an increase of decomposed bacteria. The use of oxygen in the decomposition process contributes to deoxygenation of the lake, especially in the deeper waters. Less oxygen makes it more difficult for many fish species to survive and asphyxia has become a problem in some of the deeper parts of the lake. Upwelling of deoxygenated water has caused some fish kills. Due to the deoxygenation of parts of the lake, fish are being forced to crowd into the relatively more oxygenated shallow regions, further draining the oxygen supply there. In addition, the fishing effort has traditionally been concentrated more in the shallow waters of the lake so when fish are forced into shallower waters, they also face increased risk of being caught.

Since the growth of the algae population in Lake Victoria seems to have such negative effects, we may ask whether this growth will ever cease. An arrest in growth means that the size of the population remains constant. According to this model, are there any situations in which the size of the algae population remains constant? In other words, is there any number of algae for which the population is in equilibrium?

Equilibrium is a condition in which the rate of increase equals the rate of decrease. This means that the overall rate of change is zero. Algebraically, this would be written as

$$\frac{dN}{dt} = 0.$$

Where is this true for our model? To find out, we set

$$\frac{dN}{dt} = rN$$

to 0. I.e., we consider $rN = 0$. Since r is always positive in our model, we deduce

$$N = 0.$$

The only situation for which the size of the population of algae is not changing is when there are no algae ($N = 0$). Since in this model r is always positive, for any number of algae greater than zero the population will increase exponentially. The algae population will then continue to increase indefinitely (since r is positive, $\frac{dN}{dt}$ will always be positive for this model). Thus $\frac{dN}{dt} = 0$ only occurs when there is no algae population to start with ($N_0 = 0$). The population must then remain at zero because it is not possible for algae to be created when there are not any to reproduce in the first place.

3.4. Analysis of the model

Analysis of equilibrium points sheds some light on the weaknesses of the Malthusian model. In reality, populations do not grow without bound. Mathematically, this means that a more accurate model might afford a variety of equilibrium points for the population, depending on the starting circumstances. A better model might afford other sorts of stable behavior too, such as oscillations in population size, which are also observed in nature. Nonetheless, the Malthusian model has its uses, one of which is predicting future human populations, which are assumed to be in possession of unlimited resources. Forecasters are particularly fond of making predictions about African populations. These predictions are generally based on the same model we just used for algae growth, and result in international policies concerning such things as birth control and deployment of associated medical personnel. The Malthusian model of exponential growth controls a lot of international opinion about the state of Africa (see [1] and [10] for a discussion for the changing fertility rates in Africa. Mathematically, this is discussing how the constant r is changing with time, another limitation of our model, since we assumed that r was constant).

For additional information about algae growth, including a deeply scientific explanation, see [2]. The assumption that nutrients are unlimited is not always valid. Instead of assuming algae blooms having unlimited essential nutrients, as we did in this chapter, algae are also affected by other real-world phenomena, such as the amount of light present. Light is not the defining factor in Lake Victoria, but [6] demonstrate how light can affect algae growth in other lakes, and models this process. Lake Victoria is not unique in its relations to algae blooms. For a look at another large lake, see [4], which models algae growth in Lake Erie, and explores algae biodiversity in relation to nutrients.

3.5. Differential equation terminology

A less than glamorous part of learning a new area is developing a new vocabulary. In this section we mention a lot of terminology and notation that we will use later in the book. First, the terminology for a single differential equation

(1) The phrase "differential equation" will often be denoted DE. A differential equation is an equation involving derivatives of one or more unknown functions.

(2) Dependent and independent variables: The variables you differentiate with respect to are the independent variables and the variables (or unknown functions) you are differentiating are the dependent variables. In this text the independent variable will almost always be time and the dependent variables will be the sizes of various populations.

(3) ODE and PDE: If none of the derivatives that occur in the DE are partial derivatives, the DE is said to be an ordinary differential equation or ODE (this will always happen when the unknown function depends only on a single variable in our model). If there is at least one partial derivative, the DE is said to be a partial differential equation or PDE. The DEs in this text will always be ODEs.

(4) Order: The order of a differential equation is the highest number of derivatives being applied to the dependent variable(s).

(5) Linear: A linear ODE having independent variable t and dependent variable y is an ODE of exactly the form

$$a_0(t)y^{(n)} + \cdots + a_{n-1}(t)y' + a_n(t)y = f(t)$$

for some integer n and for some given functions $a_0(t), \ldots, a_n(t)$ and $f(t)$. Here

$$y^{(n)} = y^{(n)}(t) = \frac{d^n y(t)}{dt^n},$$

the nth derivative of the unknown function y. The n is the order of the linear DE, the coefficients $a_0(t), \ldots, a_n(t)$ are the coefficients of the linear DE and the function $f(t)$ is called the forcing function.

A linear PDE is a DE which is a linear ODE in each of its independent variables.

(6) Solution: An explicit solution to an ODE having independent variable t and dependent variable x is simply a function $x(t)$ for which the differential equation is true for all values of t in some open interval. An implicit solution is an implicitly defined function (possibly multiple-valued) function which satisfies the ODE.

(7) Analytic and numerical solutions: There are theorems that guarantee the existence of a solution $x(t)$ to a DE under certain mild hypotheses (mild in the sense that functions in the real world typically satisfy them). An analytic solution, a rarity in practice, is being able to write down a general, algebraic and symbolic expression for the unknown functions. A numerical solution is typically the best we can do and is a method that gives an approximate solution to $x(t)$; i.e., given a particular $t = t_0$, a numerical solution allows for the approximation of $x(t_0)$ without knowing an expression for $x(t)$.

DE	Independent variables?	Dependent Variables?	Order	Dimension	Linear or nonlinear?
$N' = r\left(1 - \frac{N}{K}\right)N$ logistic population growth					
$C' = a - (b + d)C + eU,$ $U' = dC - eU$ model for lead in the body					
$N' = rN - bNP,$ $P' = -gP + cNP$ predator-prey model					
$S' = -bSI,$ $I' = bSI - kI,$ $R' = kI$ SIR infectious disease model					

TABLE 3.1. Some examples of differential equations.

(8) IVP: A first order initial value problem or IVP is a problem of the form

$$x' = f(t, x), \ x(a) = c,$$

where $f(t, x)$ is a given function of two variables (the independent and dependent variables, respectively, in this case) and a, c are given constants. The equation $x(a) = c$ is the initial condition and is often written IC.

Second, we introduce some terminology related to systems of differential equations. We are intentionally a little less formal in describing these terms.

(1) System of DEs: A system of differential equations is a collection of differential equations in one or more dependent variables and one or more independent variables. If none of the derivatives are partial derivatives, the system is said to be a system of ordinary differential equations.

(2) Order: The order of a system of differential equations is the highest order of the equations in the system.

(3) Dimension: The dimension of a system of differential equations is the number of equations in the system as it is written.

(4) Linear: A linear system of differential equations is a system of differential equations in which each equation in the system is linear (in all dependent variables).

(5) Solution: A solution to a system of differential equations is a collection of functions that make all the equations in the system true in an open interval. These solutions can be implicit or explicit, analytic or numerical.

(6) IVP: A system of differential equations becomes an initial value problem if it is a system of DEs and an IC for each dependent variable.

EXERCISE 3.6. Fill in Table 3.1.

3.6. Dimensional analysis and models

Units are obviously import in interpreting the model (e.g., does the population double in two seconds or in two years?).

EXERCISE 3.7. From experimental observations it is known that (up to a "satisfactory" approximation) the surface temperature of an object changes at a rate proportional to its relative temperature. That is, the difference between its temperature and the temperature of the surrounding environment. This is what is known as Newton's law of cooling. Thus, if $\Theta(t)$ is the temperature of the object after t hours, then we have

$$\frac{d\Theta}{dt} = -k(\Theta - S),$$

where S is the temperature of the surrounding environment.

Suppose that temperature is measured in degrees Celsius and time is measured in hours.

(1) What are the units of k?
(2) What sign does k have?
(3) What are equilibria for temperature? Does this make sense physically?
(4) Suppose at time 0, the temperature of the object is Θ_0. Show that

$$\Theta(t) = S + (\Theta_0 - S)e^{-kt}$$

is a solution to this IVP.
(5) Are the units in the solution in the previous part right?

EXERCISE 3.8. For the equations in Table 3.1, identify the units of all the constants in the equations.

It is also possible to use your knowledge of units and the fact that the units on both sides of an equality have to be the same to actually write down a differential equation for a model you are trying to develop. We will see examples of this later in the book.

3.7. Problems

PROBLEM 3.9. The Rule of 70 provides a simple way to calculate the approximate number of years it takes for the size of a population growing at a constant relative rate to double. In particular, it asserts that the number of years n it takes for a population growing at a rate $r\%$ is approximately,

$$n = \frac{70}{r}.$$

Justify this rule based on the material in this chapter.

PROBLEM 3.10. Consider a population of cells (e.g., bacteria) whose growth depends on the time of day. Suppose the growth rate depends (is proportional to) periodically on time with a period $\tau = 24$ hours. The simplest periodic function is $\sin t$.

(1) Write down the differential equation that describes the growth of this population. Be sure to carefully label the units of all the constants you introduce.
(2) Solve this equation analytically using separation of variables and integration.

(3) Sketch the rough shape of what the solution looks like.

(4) What are the equilibria of this system?

PROBLEM 3.11. From the model of algae growth and the rates given above, what can we predict about how large the algae population of Lake Victoria will be over a 50-year period? A range is possible, varying from the low rate of a 5-fold increase in 30 years to a high 10-fold increase in 30 years. Of course scientists have studied what algae really do in tanks and in the open. Go to the literature and see what growth rates have been observed for various algae species in various environments. Do these match the ones given in the text? How large is the error in estimating the growth constant?

The other critical constant in this model is the starting value of algae. How big an error would that estimate be likely to have? Now you can ask an important modeling question: Which error will create a bigger error in the predictive capacity of the model? To answer this question you need a way to make a comparison, both between two kinds of error and also two kinds of data collection. Furthermore you need a way to display your conclusion in a convincing graphical manner.

PROBLEM 3.12. Go online and find population data for Kenya, Tanzania and Uganda. According to the population tables for the three countries, what is the range of possibilities for population growth rates? In which of the three countries is the population growing the fastest? According to the growth rates you calculate, what will the population of Africa be 20 years from now? 50 years from now? According to the growth rates you calculate, how many years ago was the population of each of these three countries less than 100 individuals? Of course, your answer will be a range of years reflecting the range of growth rates you calculated for each country. Now, compare this answer with what the literature says about the duration of human existence in Africa. If the model is inconsistent with the literature, then what is it telling you about instead?

PROBLEM 3.13. Many people believe the population growth rate in African countries is dangerously high (see, e.g., [1, 10] and references therein). They are worried that Africa's population will outstrip its resources (especially food) in the near future, resulting in widespread starvation and disease. Other people believe that the rate of population growth one calculates from 20th century data (such as you used) does not reflect longer term rates of growth. What do you believe? How do you reconcile these two different interpretations of the Malthusian model?

Bibliography

[1] Anonymous. Africa's population: Miracle or Malthus? *The Economist*, 401 (8764):81–82, 2011.

[2] RJ Geider and Trevor Platt. A mechanistic model of photoadaptation in microalgae. *Mar. Ecol. Prog. Ser*, 30:85–92, 1986.

[3] A Getabu. Environmental factors affecting Nyanza Gulf ecosystem and fish production. In *International Conference on Sustainable Use on Biological Resources*, Budapest, Hungary, pages 26–29, August 1996.

[4] David L Howard, James I Frea, Robert M Pfister, and Patrick R Dugan. Biological nitrogen fixation in lake erie. *Science*, 169(3940):61–62, 1970.

[5] Florida LAKEWATCH. A Beginner's Guide to Water Management–Nutrients. http://lakewatch.ifas.ufl.edu/circpdffolder/Nutrintro1.pdf, 2000.

[6] M Lisi and S Totaro. Algae-light interaction: Study of an approximated model and asymptotic analysis. *Transport Theory and Statistical Physics*, 36(4-6): 323–349, 2007.

[7] Rose Mugidde. Changes in phytoplankton primary productivity and biomass in Lake Victoria (Uganda). 1993.

[8] Peter BO Ochumba and David I Kibaara. Observations on blue-green algal blooms in the open waters of Lake Victoria, Kenya. *African Journal of Ecology*, 27(1):23–34, 1989.

[9] O Ong'ang'a and K Munyirwa. Ecological threat to aquatic and Human life in and around Lake Victoria. In *Seminar for representatives of Ministry of health and public health management training institutions, Entebbe Uganda, January,* pages 27–29, 1997.

[10] Michael Walton. Are Sub-Saharan African countries in a Malthusian trap? the case of Mali. `http://www.michaelwalton.info/wp-content/uploads/2011/08/Malthusian-trap-Mali.pdf`, 2011.

CHAPTER 4

Logistic Growth

Although there are many features of Lake Victoria of interest to biologists, it is fish that we consider next. Many of the fish also live in the rivers that feed Lake Victoria and some researchers believe the fish in the lake are the result of speciation at the point where the river turns into the lake (an alternate hypothesis is that drying of the lake at some point created smaller, unconnected ponds in which isolated populations could diversify through genetic drift). The cichlids, a type of fish, in particular had a remarkable burst of speciation in response to the change from river to lake conditions. Similar speciation happened in the other lakes but in Lake Victoria, it happened much more recently, more rapidly and, due to the homogeneity of the habitat in Lake Victoria, with less opportunity for ecological isolation. This is due to the fact that cichlids are capable of rapid genetic changes and more prone to speciation than other groups of African fish [4].

Freshwater fish are an important source of protein and income to the communities in the Lake Victoria basin. Fishing in Lake Victoria is the most important economic activity. In Uganda, fishing is an important source of high quality food, employment revenue, has led to development of infrastructure, and is currently the second most important export commodity next to coffee; it generates approximately US$100 million annually. The Lake Victoria fish fauna included an endemic cichlid flock of more than 300 species. To boost fishing on the lake, Nile Perch (*Lates* spp) was introduced into the Lake in the 1950s. In the early 1980s an explosive increase of this predator was observed. Simultaneously, catches of haplochromines decreased. Although fishing had an impact on the haplochromine stocks, the main cause of their decline was predation by Nile Perch. The stocks of Nile Perch in Lake Victoria (and Lake Kyoga) were introduced from Lake Turkana and Albert, despite repeated objections from scientists [3]. This has led to the fishery being overwhelmingly dominated by three species commercially, namely mbuta (*Lates niloticus*), Nile perch in English, at 63% of the exports, dagaa/omena/mukene (*Rastrineobola argentea*), silver cyprinid in English, at 19% of exports and nyamami (*Oreochromis niloticus*), Nile tilapia in English, at 9% of the exports. Other species comprise the remaining 9%.

In several chapters in this book, we will model the fish populations. We will first try to model what the fish populations in the lake looked like before the artificial introduction of the Nile perch. In particular, we will consider the populations of seu (*Bagrus docmac* sp), a kind of catfish and fur/fuju/nkeje (*Haplochromine chichlidae*), little fish known as cichlids. We will develop a model for these populations without the presence of the Nile perch. We start in this chapter by considering the population of seu.

4.1. Biological context

Suppose we tried to use the exponential model discussed in Chapter 3 to describe the growth of seu in Lake Victoria. Using data in [10] that measures the percentage (by weight) of seu among the total fish caught by the Ugandan fisheries (a reasonable proxy for fish population), we make the plot in Figure 4.1.

EXERCISE 4.1. What are the limitations and advantages of using this proxy for fish population?

EXERCISE 4.2. Take a few minutes to look at Figure 4.1 and write down a verbal description of a function that might model the data. (E.g., it starts exponentially and then ....)

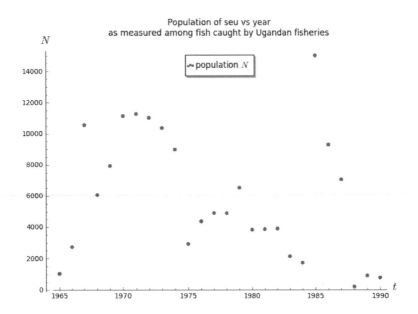

FIGURE 4.1. Percentage of seu (*Bagrus docmac* sp) among all fish caught by Uganda fisheries between 1965 ($t = 0$) and 1990 ($t = 25$). Data from [10].

The data do not appear to be terribly exponential. Suppose the data were modeled by exponential growth. If there were such a function that describes the data, it would have the form $N(t) = Ce^{rt}$ where the parameters r and c are free and determined by the data. To have any chance for solving for these two parameters we need at least two points on the curve. We choose the points $(0, 4.24)$ and $(1, 9.78)$.

EXERCISE 4.3. Using those two data points, show that $C = 4.24$ and $r \approx 0.84$.

Using the result of Exercise 4.3, we get a curve that is a poor fit to the data. See Figure 4.2 for a representation of the data and the fit we computed above. The data themselves and this fit provide pretty conclusive visual evidence that the population growth is not exponential.

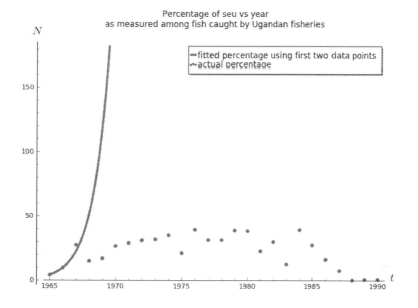

FIGURE 4.2. Percentage of seu (*Bagrus docmac* sp) among all fish caught by Uganda fisheries between 1965 ($t = 0$) and 1990 ($t = 25$). Data from [10]. Data presented along with an exponential fit for the data using the results of Exercise 4.3.

4.2. The model

In the exponential model, as time increases towards infinity, the size of the catfish population also approaches infinity. If resources are not really unlimited we might want to modify our original model to include the limitations on resources and space that exist in Lake Victoria. We might assume that there is an upper limit, called a carrying capacity (always denoted by K), on the number of catfish the lake can sustain under these limitations. One way to view the situation is that the catfish population with unlimited resources and space can increase at a rate proportional to the size of the population, as it does in the exponential model above. If there is an upper limit (K) to the amount of catfish that can live in Lake Victoria, then at any given time, some fraction of this limiting value represents available resources for growth. This fraction is the quotient of the difference between the carrying capacity and the number of catfish in the lake and the carrying capacity:

$$\frac{(K - N)}{K}$$

or, equivalently,

$$1 - \left(\frac{N}{K}\right).$$

Ecologists think of this expression as the unfulfilled potential of the lake to support catfish. It may be used to estimate the effects of the existence of a carrying capacity on the rate of change of the population. How can we modify our equation for the catfish population to include this fraction? If we multiply the rate of growth

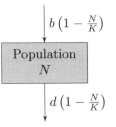

FIGURE 4.3. A box model for a population with logistic growth. This represents a population with limited resources. The population growth slows proportionally to how the fraction $\left(1 - \frac{N}{K}\right)$ of the system's resources being used by the population.

by the number of catfish that can still be sustained by the lake, we get

$$\frac{dN}{dt} = (r)(\text{ population })(\text{ room to grow })$$

$$= rN\left(1 - \left(\frac{N}{K}\right)\right).$$

Here r is the exponential growth rate for catfish found experimentally in situations where catfish have unlimited resources. If b is the rate at which catfish reproduce and d is the rate at which catfish die, we can write this as

$$\frac{dN}{dt} = (b - d)(\text{ population })(\text{ room to grow })$$

$$= bN\left(1 - \left(\frac{N}{K}\right)\right) - dN\left(1 - \left(\frac{N}{K}\right)\right).$$

In the exponential growth model we assumed that the rate of population change was proportional only to the size of the population, here we assume slightly more restrictive conditions on the rate of population growth. We assume that the growth rate and death rates depend on the size of the population.

The box model for this system is very similar to the one we had in Chapter 3. The only difference is that the rates depend on two things: the resource independent growth rate r and $\left(1 - \frac{N}{K}\right)$, what fraction of the ecosystem's resources the population is using. It can be represented visually as in Figure 4.3.

4.3. Analysis of the model

Does this new model, called the "logistic equation", fit our assumptions about carrying capacity? The expression on the right-hand side of the equation is a double proportion. That is, the rate of growth for the population is proportional to both the amount of organism present and the resource available for growth. Double either of these and you double the rate of growth. Remove either the organism or the space available and there can be no growth. This expression is just an approximation to reality, but it seems to represent more of our assumptions than simple exponential growth.

We also know that when the carrying capacity has been reached–that is, when the population of catfish is equal to the maximum sustainable population (i.e., when

$N = K$)–the size of the population should stabilize. In other words, $\frac{dN}{dt}$ ought to equal zero. For our model when $N = K$,

$$\frac{dN}{dt} = rK(1 - 1) = 0,$$

confirming our intuition that when a population is equal to its carrying capacity, that population will neither grow nor shrink.

This model seems to be a reasonable one. We have set up equations that reflect our very basic assumptions about the ecological circumstances of the catfish. Of course, there might be other equations which satisfy these assumptions too. So, if our model fails to capture the properties of the data we collect, it might not be due to wrong assumptions but to a wrong choice of mathematical model. On the other hand, our model is so simple that we might suspect that fancier equations would be saying more about our situation than we really intended to say. So, by sticking with the simplest equations that express our underlying beliefs about the system, we are putting our mental picture of the catfish-in-a-lake more fully to the test. We are asking, in effect, whether our simple assumptions about limited resources are powerful enough to predict the long term behavior of the system without having to take more subtle effects into account. We will use the mathematics of the model as a mediator between our ideas about the system and the reality offered to us by the data; we discuss this in Section 4.4.

If we solve the differential equation

$$\frac{dN}{dt} = rN\left(1 - \frac{N}{K}\right)$$

for N analytically, we can determine a function that describes the size of the population of catfish in terms of time:

(4.1)
$$N(t) = \frac{KC}{C + Ke^{-rt}}.$$

The analytic solution proceeds as follows; the basic method of solution is separation of variables.

EXERCISE 4.4. Algebraically verify the equality

(4.2)
$$\frac{dN}{N} + \frac{\frac{1}{K}\, dN}{\left(1 - \frac{N}{K}\right)} = \frac{dN}{N\left(1 - \frac{N}{K}\right)}.$$

We note that in general a complicated quotient of polynomials can be written as a sum of simple quotients of polynomials via the method of partial fractions.

Using (4.2) we get

$$\frac{dN}{dt} = rN\left(1 - \frac{N}{K}\right)$$

$$\Rightarrow \frac{dN}{N\left(1 - \frac{N}{K}\right)} = r\, dt$$

$$\Rightarrow \frac{dN}{N} + \frac{\frac{1}{K}\, dN}{\left(1 - \frac{N}{K}\right)} = r\, dt$$

$$\Rightarrow \int \frac{dN}{N} + \frac{\frac{1}{K}\, dN}{\left(1 - \frac{N}{K}\right)} = \int r\, dt$$

$$\Rightarrow \log N - \log \left| 1 - \frac{N}{K} \right| = rt + c.$$

EXERCISE 4.5. Justify the last implication either by integrating the next-to-last line to get the last line or by differentiating the last line to get the next-to-last line.

Now using some log and exponent rules we can derive:

$$\log N - \log \left| 1 - \frac{N}{K} \right| = rt + c$$

$$\Rightarrow \log \left| \frac{N}{1 - \frac{N}{K}} \right| = rt + c$$

$$\Rightarrow \left| \frac{N}{1 - \frac{N}{K}} \right| = e^{rt+c}$$

$$\Rightarrow \left| \frac{N}{1 - \frac{N}{K}} \right| = Ce^{rt}.$$

We note that the absolute values are unnecessary when N is between 0 and K since in that case $1 - \frac{N}{K} > 0$. Thus we solve for N and in this case we conclude:

$$\frac{N}{1 - \frac{N}{K}} = Ce^{rt}$$

$$\Rightarrow N = \left(1 - \frac{N}{K} \right) Ce^{rt}$$

$$\Rightarrow N = Ce^{rt} - \frac{N}{K}Ce^{rt}$$

$$\Rightarrow N + \frac{N}{K}Ce^{rt} = Ce^{rt}$$

$$\Rightarrow N \left(1 + \frac{Ce^{rt}}{K} \right) = Ce^{rt}$$

$$\Rightarrow N = \frac{Ce^{rt}}{\left(1 + \frac{Ce^{rt}}{K} \right)}.$$

EXERCISE 4.6. Finish this derivation to get the solution mentioned above in (4.1).

We can choose constants r, C, and K so that we can look at this function graphically. When we do so, we get an S-shaped curve like the one shown in Figure 4.4. We observe that for small t the graph looks exponential, and point out that this makes sense biologically.

EXERCISE 4.7. Why does it make sense biologically that for small t, the growth of a population growing logistically appears exponential? Can you also explain it mathematically in terms of the formula for $N(t)$?

This function may then be used to obtain the size of the population at any particular time or to predict long-term behavior of the population. We can try to fit a curve of this form to data, in order to deduce the values of the three constants

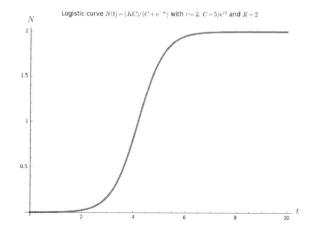

FIGURE 4.4. We see the typical S-shape of a solution to the logistic growth DE in the graph of $N(t) = \frac{KC}{C+e^{-rt}}$ with $r = 2$, $C = 5/e^{10}$ and $K = 2$.

involved and learn something about the intrinsic rate of reproduction of catfish. We could compare the shape of this family of curves to other sets of data from other systems to see how well, in general, our basic assumptions are borne out in nature.

Of course, it is a little harder to study families of curves than it is to study a single curve. When looking at ecosystems it is usually quite hard to say for sure which particular solution to a differential equation is the best stand-in for the behavior of the real system. A more cautious approach would be to make statements which are true of the set of solutions as a whole, rather than choosing a particular solution to study. If we know something about the global behavior of all, or nearly all, solutions, then we know a small error in our assignment of a particular solution to a particular situation or set of data will not result in qualitatively different predictive outcomes.

For example, we could ask whether, for all of the solutions described by our model, there are quantities of catfish for which the catfish population size is not changing. In other words, where does this model have equilibrium solutions? Once again we need to set the rate of change $\frac{dN}{dt}$ equal to zero to determine when the population of catfish (N) is constant:

$$0 = rN\left(1 - \frac{N}{K}\right).$$

Now, since r is positive, one (or both) of

$$N = 0 \text{ and } \left(1 - \frac{N}{K}\right) = 0.$$

Thus, we have

$$N = 0 \text{ or } N = K.$$

EXERCISE 4.8. Why does r have to be positive? Give biological and mathematical reasons.

Phase portraits. Of course, in this model we have an expression on the right-hand side of the differential equation that factors completely and gives all equilibrium solutions just through algebra. Most models are not so amenable to analysis. When the model is more complicated, we resort to computational methods and display the results of our computation so that features like equilibrium points are visually obvious. Figure 4.4 clearly shows a function approaching an equilibrium as time increases. But this is only one of many possible solutions to our differential equation. We get a different solution for each starting population. In order to see if all of them have the same behavior, we can analyze a slightly different graphical output: the phase portrait.

A phase portrait is a graph where time is not one of the axis variables, instead we plot points on the so-called "phase space". The dimension of the phase space of a differential equation is determined by the number of dependent variables in the differential equation. For instance, in our case, where the differential equation is

$$\frac{dN}{dt} = rN\left(1 - \frac{N}{K}\right),$$

with a single dependent variable N, we see that the phase space should be a line; typically we draw this line horizontally. To each point we attach an arrow that indicates how the dependent variable N is changing: the arrow points up if N is increasing at that population (i.e., the derivative is positive), the arrow points down if N is decreasing at that population (i.e., the derivative is negative) and there is no arrow at an equilibrium. Not only do we give the arrow a direction, we also give it a magnitude: the longer the arrow, the faster the population is growing or shrinking.

Now we consider our logistic equation. See Figure 4.5 for a pictorial representation of this process. First, we identify the domain of N and we observe that $N \geq 0$, since we cannot have a negative population. Thus, the phase portrait only consists of the solid line—we ignore points on the dashed line in Figure 4.5. Second, we identify the equilibria. Physically (and mathematically) the equilibria should be (1) when the population is zero ($N = 0$) and (2) when the population is at the carrying capacity ($N = K$). This is indicated by the two diamonds on the phase portrait in Figure 4.5. As described before, the leftmost point corresponds to $N = 0$ and the rightmost point corresponds to $N = K$. Third, we start plotting some arrows. By putting down the equilibria we have divided the phase line into two pieces. We need to pick values of N in each of those two pieces and from it deduce the direction and the size of the arrows we want to draw. In Figure 4.5 we draw the arrows to be all of the same size for ease of presentation (this is also true in most of the mathematical biology literature). In fact, though, in many other disciplines it is customary for the size of the arrow head to be determined by the size of $\frac{dN}{dt}$.

There are two cases.

(1) $0 < N < K$: i.e., N is below the carrying capacity. Biologically, this means that the population should be growing. Do we see this mathematically? Well, when $0 < N < K$ we get

$$rN\left(1 - \frac{N}{K}\right)$$

to be the product of three positive numbers: $r > 0$ because we are considering population growth, $N > 0$ because it always is, and $\left(1 - \frac{N}{K}\right)$ since $N < K$ and in particular $\frac{N}{K} < 1$. So for N between no population and the carrying capacity we see (mathematically) that the population is increasing and, therefore, the arrows should be pointing to the right (in the positive direction).

(2) $0 < K < N$: i.e., N is above the carrying capacity. Biologically this means that the population should be shrinking. See Exercise 4.9 for a mathematical explanation.

EXERCISE 4.9. By looking at the differential equation, explain mathematically why, in the case $0 < K < N$, the population is decreasing.

EXERCISE 4.10. At which N is $rN\left(1 - \frac{N}{K}\right)$ biggest? Does your answer make sense biologically?

FIGURE 4.5. A visual derivation of how to make a phase portrait for the equation $\frac{dN}{dt} = rN\left(1 - \frac{N}{K}\right)$.

We observe that the phase portrait summarizes the general behavior of a solution to the logistic equation and requires no particular knowledge of the constant K, except that it is positive. See Figure 4.6 for a side-by-side comparison of the phase portrait of the logistic equation and a general solution to the logistic equation. The dotted line is the plot of $N(t)$ vs t when the initial population is $N_0 = K$; the S-shaped curve is the plot of $N(t)$ vs t when the initial population is between 0 and K and the top curve is the plot of $N(t)$ vs t when the initial population is greater than K.

EXERCISE 4.11. What role does the sign of C play in (4.1)?

On such a graph, because t is not a variable, we do not need to have a t axis to look at the long-term behavior of solutions. Reducing the number of variables we have to plot can be useful visually. Phase graphs enable us to look at more than one solution at once, because each point on the graph represents a possible initial condition, and every path tangent to the arrows represents a possible solution to the equation. Best of all, as we will see when we start looking at systems of differential equations, they are useful for analysis of a differential equation that cannot be solved analytically, because they can be generated using numerical techniques on a good computer. The curve of a solution on a phase graph that indicates the direction of increase of time is called a *trajectory*. In Figure 4.6 the trajectory of any solution is a segment of a straight line. The figure on the right shows a couple of solutions displayed as "time series" with time as the horizontal variable, just as we are accustomed to seeing it. On the left is the phase portrait which only displays increase and decrease of N. We can see all the various long term behaviors of N at once this way, although we lose information about how fast N is increasing or decreasing.

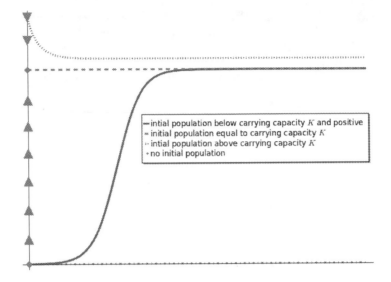

FIGURE 4.6. Three classes of possible solutions to the logistic equation, side-by-side with its phase portrait.

Stability of equilibria. Information about equilibrium solutions is particularly easy to read from a phase portrait. There are two types of equilibria: stable and unstable. Because these solutions are not changing in time, equilibria appear as points on phase graphs, and thus they are called equilibrium points or fixed points. If all the trajectories near a particular equilibrium point wind up at that equilibrium, then we say the equilibrium point is stable. After a small change to a population at a stable equilibrium, the population will return to the equilibrium or at least remain nearby (e.g., it might oscillate around the equilibrium). Instability on the other hand means that a small change in the population size will cause a major change in the system. The size of a population will not return to or even remain near its equilibrium value.

The knowledge of stability may be used to predict what will happen in a system that does not start in equilibrium. Towards which equilibrium state, if any, will the population tend? There are equilibrium values at both $N = 0$ and $N = K$. Without solving the equation, we can see what the trajectory of a solution that starts near an equilibrium point will do. Will it go towards, away from, or around the equilibrium point? If we begin with $N = K + a$, where a is a small positive quantity, the rate of change of the catfish population ($\frac{dN}{dt}$) will be negative as the phase graph indicates. This means that the population (N) is decreasing. As N decreases, a gets smaller and the rate of change of the population becomes less negative (i.e., it gets smaller in absolute value, but remains negative). The number of catfish in the population (N) continues to decrease but the rate of decrease gets smaller and approaches zero. The trajectory approaches the equilibrium solution.

If a is a small negative quantity, the rate of change of the catfish population will now be positive and the size of the population increases. As the catfish population increases, the rate of change decreases. So, as the population (N) continues to increase from a value less than K, the derivative gets smaller and approaches zero. This trajectory also approaches the equilibrium point $N = K$.

At $N = K$, then, we have a stable equilibrium solution. A small change in population around $N = K$ will result in the population increasing or decreasing slightly to return to the equilibrium solution. The size of a population of catfish will, according to this model, tend towards the carrying capacity. We can verify this mathematically by taking the limit as time approaches infinity of $N(t)$.

As long as r is positive (which we assert it is), as time elapses, N will approach K. We follow a similar approach now for the equilibrium point at $N = 0$. We start with a trajectory that begins at $N = a$, where a must be a small positive quantity (if a were negative, we would have a negative population size which, of course, is impossible). The rate of change of the catfish population will be positive as shown on the phase graph. The population (N) increases away from zero. Thus a small change from $N_0 = 0$ produces a trajectory that tends away from the equilibrium point $N = 0$.

We say that $N = 0$ is an unstable equilibrium because a small change in catfish population will cause the population to continue to grow away from that equilibrium point (rather than return to that equilibrium). This situation is illustrated if one thinks of a system in which there are no catfish to begin with ($N_0 = 0$). There is no way for this system to start miraculously producing catfish. Both N and $\frac{dN}{dt}$ stay at zero. However, if someone were to introduce even a small number of catfish, according to this model the population will grow. The population will grow large (diverging from zero) and not decrease again unless other factors are introduced. Growth will slow down as the population gets near its ceiling, K.

In Figure 4.7 we see that the catfish data from before presented with a logistic fit for the data. The fit was found in almost the most naive way possible. The process is described in Exercise 4.12.

EXERCISE 4.12. We claim that in trying to fit the data in Figure 4.1 we can assume that $K = 39.4$. Why is that a reasonable assumption? Then, using the two points $(0, 4.24)$ and $(1, 9.78)$ we can find that $r \approx 1$ and $C \approx 4.77$.

4.4. Model versus real world data

We can compare our graph of the catfish data to a choice of curve for $N(t)$. We can clearly see that there is still quite a bit of error in our model. One of the main reasons for this may have to do with the fact that the cichlids, the main source of food for catfish, decreased in number over this period of time. Not only is there a limited amount of resources available but that amount decreases; it is not constant as our model assumes.

In fact, to see how far from a logistic or exponential fit we might have, we calculate the least squares fit for a logistic and exponential regression[1] in Sage. We get the results of Figure 4.8 which are both visually pretty terrible.

---

[1]A least squares regression of observed data to a theoretical curve (e.g., linear, exponential, logistics, etc.) is a method to find that curve that minimizes the error between the data and the curve by choosing parameters that determine the curve.

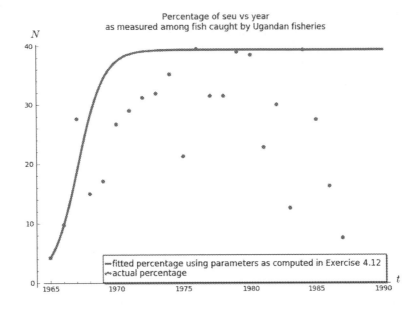

FIGURE 4.7. Percentage of seu (*Bagrus docmac* sp) among all fish
caught by Uganda fisheries between 1965 ($t = 0$) and 1990 ($t = 25$).
Data from [10]. Data presented along with a logistic fit for the data
using the results of Exercise 4.12.

4.5. Implications of the model

The assumptions that we made in this chapter about the population of seu are
clearly flawed. As mentioned above, a logistic model assumes that the growth rate
of the seu population is proportional to how many seu there are and to how far away
they are from the ecosystem's carrying capacity for seu. The seu live in a much
more complicated system than that. There are other things that have an effect on
the number of seu. The introduction of predators (e.g., the nile perch) and the
extinction of prey (e.g., the cichlids) will clearly have an effect on the population of
seu. In particular, one can think of this as saying that the carrying capacity itself is
a function of time whereas in our logistic model we have assumed it was constant.
See [1] for a rich model of an analogous ecosystem in Florida and [5, 6, 8, 9] for a
thorough discussion of how to model the fish population in Lake Victoria. We will
return to this kind of modeling in Part 4.

4.6. Problems

PROBLEM 4.13. Do you believe that the disagreement between the data on the
Bagrus catfish and the logistic model is a result of a decrease in prey fish, as is
claimed in this chapter? The fall-off of the percentage of catfish caught is probably
an indication of lower population, but what about the sudden increase in percent
of catch? Could this be explained by other factors? Analyze these factors, decide
which are likely to have the most prominent effects, and then describe how you
might modify the model for catfish population as a result. Catch data for Bagrus

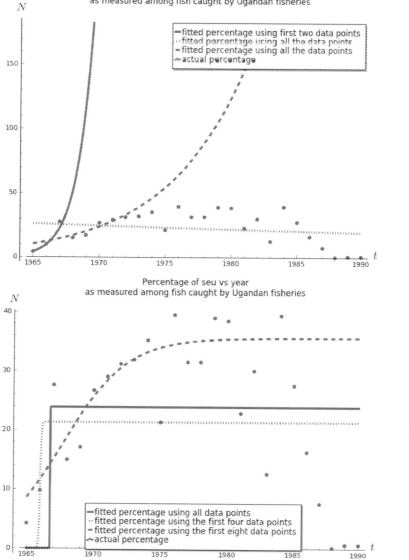

FIGURE 4.8. Various exponential regressions and logistic regressions for the data in Figure 4.1 were computed using Sage.

catfish, tilapia, cichlid and other fish in Lake Victoria can be found online. How does the logistic model fit with these data? How about your refined model?

PROBLEM 4.14. Suppose you have two populations living next to each other, each increasing in population according to the logistic equation. In addition, every year there is a small percentage of those in population A who will become a member of population B, and *vice-versa*. Each population's loss is a gain for the other. How

Year	Period	Population	$\left(1 - \frac{N}{K}\right)$	$\frac{dN}{dt}$
1905	1	10		
1930	2	613		
1935		1189		
1940		2139		
1944	3	3157		
1946		3750		
1950		4810		
1996	4	7500		

TABLE 4.1. Elephant population in Kruger National Park based on a model that assumes a carrying capacity K of 7500 elephants and an intrinsic per capita growth rate of 0.15 yr^{-1}.

would you model the change of population in both populations? What equations would govern the system? What, if any, are the equilibria?

PROBLEM 4.15. In the early 20th century, Kruger National Park was established in the Northeastern part of South Africa (see [2] for more details on the history of the park and the management of its elephant population). In 1905, two years after the park was established, there were 10 known elephants in the park. Up until the 1960s a preservationist approach was taken in which the elephants were permitted to grow without limitation. Between the 1960s and the mid 1990s the elephant population was artificially limited to about 7000 via the process of elephant culling because too many elephants in one place led to a variety of problems, both for elephants and the environment. In this problem we will model the elephant population in the park up to the mid 1990s. The population grew over four time periods: first it was mostly empty and some elephants migrated in from surrounding areas, second it was a game reserve, third the population was left to grow on its own and fourth population culling took place.

Table 4.1 is based on a model with $r = 0.15$ yr^{-1} and $K = 7500$ elephants. This model is based on data available in [11]. Table 4.2 is the populations as reported by the wardens of the park. These data were found in [11].

Now, do the following:

(1) Fill in Table 4.1.
(2) We see that the maximal per capita birth rate is $r = 0.15$ yr^{-1}. In which period is the effective per capita birth rate closest to the maximal one? In which period is the effective per capita birth rate the smallest?
(3) For each time period, what is the relationship between the elephant population's growth rate and the unused portion of K?

Year	Period	Population	Year	Period	Population	Total culled
1903	1	0	1967	4	6586	355
1905		10	1968		7701	460
1908		25	1969		8312	1160
1925	2	100	1970		8821	1846
1931		135	1971		7916	602
1932		170	1972		7611	608
1933		200	1973		7965	732
1936		250	1974		7702	764
1937		400	1975		7408	567
1946	3	450	1976		7275	285
1947		560	1977		7715	544
1954		740	1978		7478	348
1957		1000	1980		7454	356
1960		1186	1981		7343	16
1962		1750	1982		8051	427
1964		2374	1983		8678	1290
			1984		8273	1289
			1985		6887	268
			1986		7617	404
			1987		6898	245
			1988		7344	273
			1989		7468	281
			1990		7287	232
			1991		7470	218
			1992		7632	185
			1993		7834	308
			1994		7806	177

TABLE 4.2. The population of elephants in Kruger National Park as calculated by the wardens.

(4) At what time was the growth rate biggest? Is that consistent with what a solution to this differential equation looks like?

(5) Describe numerically what happens to the entries in the table as the population of elephants approaches K.

(6) Make a scatter plot of the data in Table 4.2 on the same axes as the data from the model. What do you observe?

(7) Recall that the assumptions of the logistic model include that population growth is continuous and there are no time lags. What makes these assumptions not exactly true in nature?

(8) On your scatter plot, sketch, by hand, what the curve we use to model this system would look like if the model took into account that crowding effects are delayed by several years?

This problem is based on [7].

Bibliography

[1] Donald L DeAngelis, William F Loftus, Joel C Trexler, and Robert E Ulano-
    wicz. Modeling fish dynamics and effects of stress in a hydrologically pulsed
    ecosystem. *Journal of Aquatic Ecosystem Stress and Recovery*, 6(1):1–13, 1997.

[2] SM Ferreira, S Freitag-Ronaldson, D Pienaar, and H Hendriks. Elephant ma-
    nagement plan: Kruger National Park. *Pretoria, SANParks*, 2011.

[3] Geoffrey Fryer. Concerning the proposed introduction of Nile perch into Lake
    Victoria. *East African Agricultural Journal*, 25(4):267–70, 1960.

[4] GEF. Kenya, Tanzania, Uganda - Lake Victoria Environmental Management
    Project.   http://documents.worldbank.org/curated/en/1996/06/696464/
    kenya-tanzania-uganda-lake-victoria-environmental-management-project,
    1996.

[5] JYT Mugisha and H Ddumba. Modelling the effect of Nile perch predation and
    harvesting on fisheries dynamics of Lake Victoria. *African Journal of Ecology*,
    45(2):149–155, 2007.

[6] M Njiru, E Waithaka, M Muchiri, M Van Knaap, and IG Cowx. Exotic intro-
    ductions to the fishery of Lake Victoria: What are the management options?
    Lakes & Reservoirs: Research & Management, 10(3):147–155, 2005.

[7] University of Wisconsin Board of Regents. Population dynamics. `http://
    ats.doit.wisc.edu/biology/ec/pd/pd.htm`, 2003.

[8] Richard Ogutu-Ohwayo. The decline of the native fishes of lakes Victoria and
    Kyoga (East Africa) and the impact of introduced species, especially the Nile
    perch, *Lates niloticus*, and the Nile tilapia, *Oreochromis niloticus*. *Environ-
    mental biology of fishes*, 27(2):81–96, 1990.

[9] Richard Ogutu-Ohwayo. The effects of predation by Nile perch, Lates nilo-
    ticus L., on the fish of Lake Nabugabo, with suggestions for conservation of
    endangered endemic cichlids. *Conservation Biology*, 7(3):701–711, 1993.

[10] FL Orach-Meza. Present Status of the Uganda Sector of Lake Victoria Fisher-
    ies, 1992.

[11] Ian J Whyte, Rudi J van Aarde, and Stuart L Pimm. Kruger's elephant po-
    pulation: its size and consequences for ecosystem heterogeneity. *The Kruger
    experience: ecology and management of savanna heterogeneity. Island, Wa-
    shington, DC*, pages 332–348, 2003.

CHAPTER 5

Harvesting a Population with Logistic Growth

The logistic equation we just studied is a far better model of how organisms grow than exponential growth was, unless the population is at the very start of its growth pattern. Even though the fit to Bagrus catch data was not very good, the logistic model captured one very important aspect of it: not growing without bound. No species can grow without bound since in every ecosystem there are limited resources. The logistic model therefore has another advantage: it builds in the effect of limited resources explicitly, so that the mathematical consequence is causally tied to our hypothesis of what causes it. We always seek this elegance in a model. We want our hypotheses clearly expressed in the equations we create, and we want the solutions to those equations, which illustrate the consequences of our hypotheses, to match our observations of nature.

Mathematically, here is a way to think about the Bagrus data. If we compare the exponential model to the data, the error in predicting a data point from the model gets larger and larger as time goes on. The exponential function keeps increasing but the data are bounded, so the error goes to infinity with time. The logistic curve, being bounded itself, results in an error that, although large, is bounded over time. One must admit that the improvement from an infinite error to a finite one is impressive, even though the finite error may (as in this case) be very large and the fit-to-data is not at all convincing.

Populations that really do grow in limited environments with no other forces acting on them do indeed obey the logistic equation fairly closely. It may be hard to imagine such a situation in nature but it is easy to imagine setting one up in a laboratory. Small organisms such as bacteria or insects that have a constant source of nutrients can actually be observed to grow according to the logistic equation, as in [5, 2, 3]. In nature, more complicated things usually happen. When we try to model these complicated things, a good strategy is to go after basic qualitative observations before attempting to match numerical data. In this way we can systematically test the validity of our assumptions of causality by building them into the equations we are using, teasing out which assumptions are required for a certain qualitative phenomenon to occur. The idea of a limit to growth is an example of such an observed phenomenon (in reverse, as nobody has ever seen an example of unlimited growth). The corresponding idea of limited resources is the causal explanation built into the logistic equation.

5.1. Biological context

Let us look now at another observed phenomenon: extinction. The cichlids of Lake Victoria are not a single species but many. Although related through an-cestry and evolution, these species maintain their current distinctions by inhabiting different niches. Differing diets and distinct habitat preferences keep cichlid populations separate, allowing coexistence of many seemingly similar species. However, over the course of the last century many species of cichlid have gone extinct in Lake Victoria. These species, if they currently exist, live in household aquariums around the world, propagated and disbursed by fish enthusiasts, but they are gone from their birthplace.

Species can go extinct and do so regularly. Ecologists offer a variety of hypotheses when this happens. The extinction of a species could be the result of direct human interference such as habitat loss or overfishing, or "natural" causes that may be influenced directly by humans or the larger environment, such as introduction of new species or habitat change due to weather patterns. Extinctions, though they have occurred throughout the history of life on earth, have come to be considered a bad thing because so often the cause is attributed to human interference and because the incidence of extinction appears to have increased. It is important to note that under some circumstances, extinction is desirable. For example, if the organism is a human disease such as smallpox and the habitat for that organism is the human population itself, then extinction of the organism means eradication of a human disease. If the organism is a single infection inside a person, then extinction of the species (at least in its immediate habitat) means the disease is cured. These examples also illustrate that ecological models and medical models are closely related, the difference often being just that of context and interpretation. We will see more about medical models in Parts 3 and 4.

Let us take the example of fishing. What does it mean for a species to be fished to extinction? The possibility of fishing a species to extinction is now discussed regularly with regard to valuable commercial varieties, such as the Atlantic cod and in Lake Victoria, the Nile Perch. Does "fishing to extinction" mean one would have to catch every last cod in the Atlantic Ocean? This seems like an improbable requirement. Once the cod population drops below a commercially viable number, fishermen will no longer go to the trouble to catch them. Big fishing operations catch species indiscriminately but even they will not be economically viable if fish populations drop enough. Moreover, it might be possible to fish a species to extinction by reducing the population of the species to the point where it cannot, for whatever reason, procreate faster that it dies off. Catching every individual (or mating pair) might not be necessary.

In the 1950s Warder Allee pointed out the existence of such an effect in natural populations, which has since come to be known as the Allee effect. Researchers have proposed many reasons for this effect, depending on the species under consideration. Often the difficulty of locating mates at low population densities is cited, or in some cases the need for many adults in the rearing of broods. Predator behavior patterns are also a possible justification. Any model that needs to address questions of extinction must display some kind of Allee effect at low population sizes. Ecologists sometimes refer to this as an "extinction threshold." Their guess is that there is some cutoff for population size. If the population were to fall below that number, it would automatically go extinct (usually for one of the reasons cited above). In terms of modeling, we would say that we expect the hypothesis (difficulty of locating mates, for example), which we would build into the equations, would lead to a conclusion (the extinction threshold, usually demonstrated graphically as a phase portrait).

5.2. The model

Let us look once again at the logistic equation $X' = kX(1 - X)$ and its phase portrait in Figure 5.1. Here we have renormalized the variables in such a way that the carrying capacity is 1.

FIGURE 5.1. The phase portrait for the logistic equation $X' = kX(1 - X)$.

It is easy to see from the phase portrait that, no matter how we adjust this organism's inherent growth rate and the carrying capacity of the environment, this model will never display an Allee effect.

EXERCISE 5.1. Explain the above comment. What is the "easy" observation?

No matter how small the starting population is, as long as it is positive the population will rise to a stable equilibrium. This model, although excellent at reproducing observed limits to growth, is unable to reproduce observed extinctions at low population sizes. Here is another way to see this constraint:

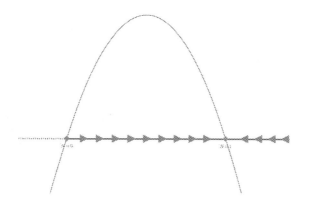

FIGURE 5.2. The phase portrait for the logistic equation $X' = kX(1 - x)$. The function $y = kX(1 - X)$ is superimposed upon the phase portrait. Between $X = 0$ and $X = 1$ we see that $y = X'$ is always positive and so X is always increasing if X is between 0 and 1.

The function describing the differential equation for logistic growth is always positive between zero and the carrying capacity. At no point in that interval can the population drop at all. To reproduce an Allee effect we might want to see a function that looks more like the phaseportrait in Figure 5.3.

So, from a modeling standpoint, the question is whether any of the reasons offered for the Allee effect as observed in nature will result in a picture like the one in Figure 5.3, which gives the desired qualitative outcome of an extinction

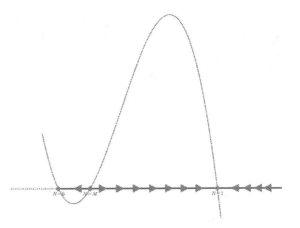

FIGURE 5.3. A phase portrait of a differential equation that encodes the Allee effect. In particular, we see that if the population X is below some threshold M we see that $X' < 0$ and so the population $X \to 0$.

threshold below which the population goes to zero. The extinction threshold is also an equilibrium of the system, but an unstable one. Above the threshold, the population rises to some stable value. Below it the population drops to zero.

Notice that the phase portrait displays the unstable equilibrium clearly, whereas one cannot observe it by graphing solutions against time. Unless the initial value is exactly at the critical value the solution will shoot up or down. Even if it begins at the critical value, roundoff error may be enough to push the solution towards a different equilibrium.

Now let us look at several hypotheses that might account for the existence of an Allee effect in a population. Suppose we start with a population that obeys logistic growth (with instrinsic growth rate r) and add a death rate due to predation. Our equation will have this form:

$$(5.1) \qquad X' = rX(1 - X) - \text{(death due to predation)}.$$

EXERCISE 5.2. What should the units be of term being subtracted?

Let us consider two options for predator behavior. The predator could take a constant proportion of prey per unit time. This assumption would result in an equation like this:

$$(5.2) \qquad X' = aX(1 - X) - d_P X.$$

EXERCISE 5.3. What are the units of d_p? What is the physical/biological meaning of d_p?

EXERCISE 5.4. Show that the equilibria of this model are $X = 0$ and $X = (r - d_p)/r$.

Solving for equilibrium values in this model as in Exercise 5.4 yields two equilibria, one at zero and one positive value at $X = (r - d_p)/r$. Between these equilibria the function describing X' is positive. The phase portrait looks as in Figure 5.4.

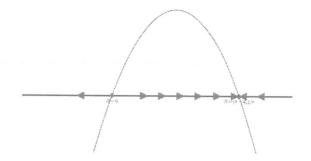

FIGURE 5.4. Phase portrait of the logistic model with proportional harvesting (predation): $X' = rX(1 - X) - d_pX$. Superimposed upon the phase portrait is the graph $y = rX(1 - X) - d_pX$.

Qualitatively there is no difference between the behavior of this system and that of the original logistic equation that we modified. The population stabilizes at a value slightly below 1 and no Allee effect is present.

EXERCISE 5.5. Explain why there is no Allee effect present in this model. Explain it both in terms of the phase portrait and the parabola superimposed upon the phase portrait as in Figure 5.4.

Since this model does not possess the Allee effect, we reject it try to find a better model.

Now we assume the predator always takes a constant amount of prey. In that case the equation would look like this:

(5.3) $$X' = rX(1 - X) - d_p.$$

Note that this is unambiguously *not* the same d_p as in the model $X' = rX(1 - X) - d_pX$.

EXERCISE 5.6. What are the units of d_p? What is the physical meaning of d_p?

5.3. Analysis of the model

We consider the model

$$X' = rX(1 - X) - d_p.$$

In this case there are also two equilibrium values at $X = 1/2 \pm \frac{\sqrt{1-4d_p/r}}{2}$.

EXERCISE 5.7. Show that the two equilibria of this model are at $X = 1/2 \pm \frac{\sqrt{1-4d_p/r}}{2}$.

The phase portrait for (5.3) can be seen in Figure 5.5.

The lower equilibrium is unstable and functions as an extinction threshold. Below this value the population dies off. Above it the population rises to a stable equilibrium value. As long as $r > d_p > 0$ we see an Allee effect in this model. As d_p gets small the region below the extinction threshold shrinks and the model looks more and more like the logistic equation.

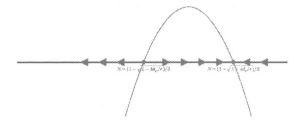

FIGURE 5.5. Phase portrait of the logistic model with constant harvesting (predation): $X' = rX(1-X) - d_p$. Superimposed upon the phase portrait is the graph $y = rX(1-X) - d_p$.

EXERCISE 5.8. Justify these claims:

(1) As long as $r > d_p > 0$ we see an Allee effect in this model.
(2) As d_p gets small the region below the extinction threshold shrinks and the model looks more and more like the logistic equation.

There is one problem, however. In the logistic equation, $X = 0$ was a stable equilibrium. This makes intuitive sense for populations. We have lost this nice feature in (5.3). At $X = 0$ the growth rate is actually negative, pushing the population into negative numbers, a biologically impossible situation. The model itself becomes an unreasonable description of predator behavior as the predator attempts to consume more prey than exists.

A modeler eager to preserve the qualitative feature of the Allee effect would have to make an aesthetic decision at this point. One could use this model to study the effect with the overt recognition that it is only valid for $X > 0$. In this case it could not be used to study what might happen at very small population sizes. Alternatively a modeler could reformulate the death rate as a function that is roughly proportional to X for small values of X but approaches a constant value as X gets large. Models incorporating this feature are in the literature, as are models that consider mating difficulties and other hypotheses.

5.4. Implications of the model

The demise of the cichlid species of Lake Victoria is often attributed to the mid 20th century introduction of the Nile Perch, a wide ranging general predator. Since the introduction of this species, a large number of cichlid species have disappeared from the lake. Models such as the ones in this chapter are consistent with the hypothesis that cichlid extinctions are linked to the introduction of this fish. This event has been written about at great length: for example, see [6, 7, 8]. But the devil is in the details. Top predators can also be "keystone species" that keep competing populations in check and whose removal results in extinctions. The models in this text address both situations but which situation prevails depends on the details of the relationships among species. The story of the Nile perch is not over.

5.5. Problems

PROBLEM 5.9. Let N be the population of a species at times t. The continuous population growth models are of the form:

$$\frac{dN}{dt} = \text{births} - \text{deaths} + \text{migration}.$$

Migration is liberally interpreted to mean, e.g., culling, harvesting, hunting, predation, etc.

(1) Explain how the exponential growth model is consistent with this idea.
(2) Recall the logistic growth model that we saw in detail in the previous chapter:

$$\frac{dN}{dt} = rN\left(1 - \frac{N}{K}\right).$$

Is this consistent with the same idea? Why or why not?
(3) What are the units of $\left(1 - \frac{N}{K}\right)$? What is the meaning of that factor (hint: consider the cases when $N = 0$ and $N = K$)?
(4) Say we introduce predation, a special kind of migration. Let $p(N)$ be the rate of predation. What would the differential equation for $\frac{dN}{dt}$ be now?
(5) Sketch a graph of what you think $p(N)$ might look like.
(6) A particular form of $p(N)$ proposed by Ludwig *et al.* [5] is $\frac{BN^2}{A^2 + N^2}$. What are the units of the various parameters? What is the physical meaning of A (try plotting a few of these curves with varying values of A)? What is the differential equation now?
(7) Choose some reasonable values for the constants in the models you have developed up to now and numerically find the equilibria. Then determine what happens if you start a little bit above or below each equilibrium (if it makes sense to be below or above the equilibrium).

PROBLEM 5.10. This problem is based on [4]. We have reworked it for the elephant population in Kruger park as in Chapter 4.

(1) For the elephant population in Kruger park without harvesting, $N_0 = 1186$ elephants, $K = 7500$ elephants, $r = 0.15$ per year. Try introducing some harvesting by changing the harvesting percentage to 5% or $b = 0.05$. What happens to the population over 10 years? Reduce the harvesting to 1%. What happens?
(2) What is the smallest percentage harvesting that you can have that will cause extinction in this species? You may need to increase the stop time so that you can see the long term trend. Give your answer in percentages to one decimal place. Using this harvesting percentage increase the value of r. What happens? Find the smallest percentage harvesting that you can have that will cause extinction with this species. Repeat this a few times. What trend do you see? How would you explain it?
(3) In 1980, 356 elephants were culled, with previous years recording similar numbers. If the park were to target at most 500 elephants to be culled every year, can the deer population be maintained?
(4) What is the maximum percentage culling that the park can allow without the population decreasing below 3,000?

PROBLEM 5.11. In [1] a distinction is made between *strong* and *weak* Allee effects. What we define in this section is a strong Allee effect: namely, one that gives rise to a critical population size below which the population will go extinct. The authors of [1] say

> the difference between Allee effects and the classical (Malthusian) logistic theory is seen most clearly in the correlation between per capita growth rate and population size.

(1) The classical logistic model is given by $N' = rN(1 - N/K)$ where N is the population, K is the carrying capacity and r is the intrinsic birth rate of the population. Show that the *per capita* growth rate does not increase with population size.

(2) An equation such as

$$\frac{dN}{dt} = rN \left(1 - \frac{N}{K}\right) \left(\frac{N-a}{K}\right),$$

the authors claim, can be used to illustrate a strong Allee effect.

   (a) What are the units of a? Do the units of the two sides match up?

   (b) What are the equilibria? What does a represent?

   (c) Draw the phase portrait of this model. Superimpose the function $y = N'$ as we did in the figures in this chapter.

   (d) What can be said about the *per capita* growth rate of the population?

Bibliography

[1] JM Drake and AM Kramer. Allee effects. *Nat. Educ. Knowl.*, 3(10):2, 2011.

[2] H Einarsson and SG Eriksson. Microbial growth models for prediction of shelf life of chilled meat. *Science et Technique du Froid (France)*, 1986.

[3] Angela M Gibson, N Bratchell, and TA Roberts. The effect of sodium chloride and temperature on the rate and extent of growth of *Clostridium botulinum* type A in pasteurized pork slurry. *Journal of Applied Bacteriology*, 62(6):479–490, 1987.

[4] Brandon M. Hale and Maeve L. McCarthy. An Introduction to Population Ecology. *Convergence*, 2005.

[5] D Ludwig, DG Aronson, and HF Weinberger. Spatial patterning of the spruce budworm. *Journal of Mathematical Biology*, 8(3):217–258, 1979.

[6] Ole Seehausen, Frans Witte, Egid F Katunzi, Jan Smits, and Niels Bouton. Patterns of the remnant cichlid fauna in southern Lake Victoria. *Conservation Biology*, pages 890–904, 1997.

[7] Frans Witte, Tijs Goldschmidt, Jan Wanink, Martien van Oijen, Kees Goudswaard, Els Witte-Maas, and Niels Bouton. The destruction of an endemic species flock: quantitative data on the decline of the haplochromine cichlids of Lake Victoria. *Environmental biology of fishes*, 34(1):1–28, 1992.

[8] Frans Witte, BS Msuku, JH Wanink, O Seehausen, EFB Katunzi, PC Goudswaard, and T Goldschmidt. Recovery of cichlid species in Lake Victoria: an examination of factors leading to differential extinction. *Reviews in Fish Biology and Fisheries*, 10(2):233–241, 2000.

CHAPTER 6

Euler's Method

Up until now in this book we have been considering ordinary linear differential equations (actually initial value problems) of the form

$$\frac{dN}{dt} = f(t, N(t))$$
$$N(a) = c.$$

The equations that we have examined up to now are summarized in Table 6.1 and our interest is in IVPs for these differential equations. In only some of these cases can the particular solution to the IVP be found analytically. We will consider numerical solutions to these IVPs in most of the cases.

What does it mean to have a numerical solution to an IVP? This means that we will not always be able to find an expression for $N(t)$ in terms of functions like $\sin t$, $\ln(t)$, etc. The best we can hope for is that we discover a way to approximate $N(t_0)$ up to some known error for every t_0 for which a solution would be defined.

How is this done? There are several ways to do this. We refer the reader to any standard elementary book on differential equations for more details on Euler's method and other numerical methods for solving differential equations (e.g., [1, 2]).

6.1. The method

Like much of the very best mathematics, the method we describe below is due to Euler and proceeds as follows (there are other more efficient and effective methods for numerically solving IVPs, but we focus on Euler's method because of how clear

Model	$f(t, N(t))$	Location in text
Tumor growth	$rN(t)\ln\left(\frac{K}{N(t)}\right)$	Chapter 2
Exponential growth	$rN(t)$	Chapter 3
Logistic growth	$rN(t)\left(1 - \frac{N(t)}{K}\right)$	Chapter 4
Logistic growth with linear harvesting	$rN(t)\left(1 - \frac{N(t)}{K}\right) - bN$	Chapter 5
Logistic growth with constant harvesting	$rN(t)\left(1 - \frac{N(t)}{K}\right) - b$	Chapter 5
Logistic growth with Allee effect	$rN(t)\left(1 - \frac{N(t)}{K}\right)\left(\frac{N(t)-a}{K}\right)$	Chapter 5

TABLE 6.1. Summary of the differential equations seen up to now.

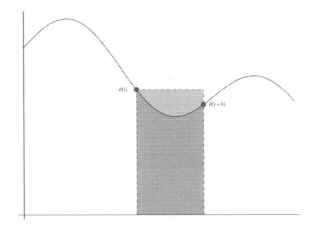

FIGURE 6.1. The integral that appears in Euler's method and an approximation of the integral.

it is). We start by trying to integrate both sides:

$$\frac{dN}{dt} = f(t, N(t))$$
$$\Rightarrow dN = f(t, N(t))\ dt$$
$$\Rightarrow \int_t^{t+h} dN = \int_t^{t+h} f(t, N(t))\ dt$$
$$\Rightarrow N(t+h) - N(t) = \int_t^{t+h} f(t, N(t))\ dt$$
$$\Rightarrow N(t+h) = N(t) + \int_t^{t+h} f(t, N(t))\ dt.$$

So, what does this tell us? It tells us that if we can estimate this integral then we determine the size of a population at a time h units in the future only knowing the size of the population at time t.

What is the easiest possible approximation to this integral? Consider Figure 6.1. We see the graph of a function $N(t)$ between t and $t + h$. The darker region in the figure is exactly the integral

$$\int_t^{t+h} f(t, N(t))\ dt.$$

An easy estimate for this integral is the area of the rectangle of width h and height $f(t, N(t))$, represented in the figure as the region within the dashed lines. What is the area of that rectangle? Well it is nothing more than $hf(t, N(t))$.

This leads us to Euler's method:

$$N(t+h) \approx N(t) + hf(t, N(t)).$$

EXERCISE 6.1. Using Euler's method, complete the following tables for

$$N' = 0.05 \left(1 - \frac{N}{100}\right) \left(\frac{N-20}{100}\right)$$

t	$N(t)$	$N'(t)$
0.0	21	
0.1		
0.2		

t	$N(t)$	$N'(t)$
0.0	19	
0.1		
0.2		

6.2. Problems

PROBLEM 6.2. Consider the following theorem: Suppose the general first order ordinary differential equation: $N' = F(t, N)$, $N(t_0) = N_0$. Suppose that $F(t, N)$ is a continuous function defined on some rectangle:

$$R = \{(t, N) : t_0 - \delta < t < t_0 + \delta, N_0 - \epsilon < N < N_0 + \epsilon\}.$$

Then there exists some number δ_1 (possibly smaller than δ) so that a solution $N = f(t)$ to the differential equation is uniquely defined for $t_0 - \delta_1 < t < t_0 + \delta_1$. Explain what this says about when Euler's method is guaranteed to work.

PROBLEM 6.3. Consider the differential equation $N' = t^2 + N^2$. Figure 6.2 shows a *slope field* for this differential equation. Given an ODE $N' = F(t, N)$ to each point (t_0, N_0) we attach a little of slope $F(t_0, N_0)$. The solution to the ODE through the point (t_0, N_0) must be tangent to the little line segment we drew.

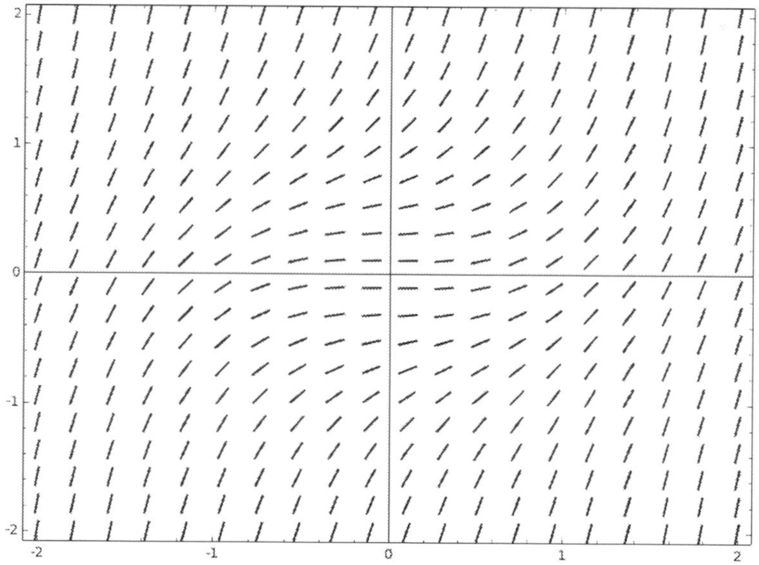

FIGURE 6.2. A slope field for the differential equation $y' = t^2 + y^2$.

(1) For the particular solution through the point $(-1, 0)$, estimate the value $N(0)$ using the slope field.
(2) What's the relationship between the slope field and Euler's method?

Bibliography

[1] William E Boyce, Richard C DiPrima, and Charles W Haines. *Elementary Differential Equations and Boundary Value Problems*, volume 9. Wiley, New York, 1992.

[2] William F Trench. *Elementary Differential Equations with Boundary Value Problems*. Brooks/Cole Thomson Learning, 2013.

CHAPTER 7

Modeling Interlude: The Modeling Process

Up to now we have seen several models: the Gompertz model for tumor growth, Malthusian population growth (growth with unlimited resources), logistic population growth (growth with limited resources) and logistic population growth with harvesting. These models are classic and have been around a while. The development of these analytic models deepened our understanding of the systems for which they were developed. In this interlude, we step back and take a broader perspective to the modeling process.

We assume that you are reading this book because you want to understand how to model biological systems and processes using mathematics. In this book we focus exclusively on models that can be understood using continuous mathematics: in particular, we focus on systems and processes that can be described by systems of differential equations. Therefore, although what we present in this chapter is general advice that can be applied to other kinds of models, we illustrate everything using the kinds of models and mathematics we discuss in the text. In a sense, all we are doing is breaking down the scientific method, combining it with mathematical problem posing and solving and re-packaging them for you.

7.1. Added value

Before we get to particular advice, we should mention what the minimum requirement is for a model to be useful. Fundamentally, the goal of *all* modeling problems is added value. That is, we use all our knowledge of the system we are attempting to model, develop the model, run the model (maybe many times) and collect the resulting data. The added value of the model is the construction of new knowledge. Three kinds of added value the models in this book could give us are: (1) they can summarize data, (2) they can explore mechanisms and (3) they can make predictions. These three items are used in order to test hypotheses, the main activity with which scientists engage.

We illustrate these three kinds of added value by considering the following scenarios that arise in pharmacokinetics (the study of how drugs move through a patient's body). We will discuss several pharmacokinetic models and the basic techniques and terminology of the discipline in Part 3 but we think that merely having an informal understanding of how such a model might look is enough to understand and appreciate the advice we give here.

Suppose we want to understand how drug X moves through a patient's body. A standard box (or compartment) model in pharmacokinetics assigns a box to each of the body's systems that might have something to do with drug X. For example, if the drug is taken orally, there might be a box for the stomach. The drug eventually travels through the body and then out through the kidneys. Maybe drug X is targeting a sore throat. How will it get there? How much will get there? Pharmacokinetics asks questions like these and uses a box model to do so.

A model designer needs to know enough biology to propose a reasonable collection of boxes and the ways they are connected. For example the connection from stomach to blood stream is almost always one-way: from the stomach to the blood stream. Some connections, however, are two-way: in Chapter 12 you will see that airborne lead enters a child's bloodstream through their lungs and is absorbed by the bone. With time, the accumulated lead in the bone will leech back into their blood. Once a reasonable model is visualized, the model-designer needs to know enough mathematics and biology to know at what rate the drug goes from box to box. This requires work: e.g., an extensive literature search, fit to a good data set, lab work or, as a last resort, creative guessing. Finally, the model-designer needs to know enough mathematics to be able to analyze the model and correctly interpret the results.

With this background, we now highlight three ways in which a model can provide added value. We consider the particular problem in pharmacokinetics that asks: "There is a drug X and a group of patients P. What should the dosing regimen for drug X for a patient who is like patients in group P so that the drug will be as effective and safe as possible?"

- **Make predictions:** Once a satisfactory model and parameter values have been determined, we can make predictions about the system we are modeling. For example, can we accurately predict the concentration of drug X in a patient's blood 6 hours after administration of the drug?
- **Explore mechanisms:** Developing models is an important step in understanding how biological systems function. After developing good models it is possible to explore correlations between model parameter values and empirical parameters. For example, in our model we hypothesize that drug X is absorbed by the bone and then re-released into the blood but the data are utterly different than what we get with our model. Is our understanding of how drug X is released into the blood fundamentally wrong?
- **Summarize data:** Biological models are very useful for summarizing data. A suitable model with good parameter value estimates and estimates of their uncertainty can be helpful. For example, say our model tells us that a patient who weighs 150 lbs $\pm$ 10 lbs should receive a dose of 10 mg in order for drug X to be effective. Then, this single model developed with a handful of patients contains a range of information that can be used with other patients.

We hope that these three examples highlight the usefulness of having models and what kind of new knowledge can be constructed and extracted by the proper use of a mathematical model.

7.2. A list of questions for a model-designer to consider

In this book we are considering box models but this next list of questions are ones that can be asked by any person modeling anything mathematically. They are general questions that focus your attention on the information that you need for what you want to do.

(1) **Identify the problems and questions.** What do you wonder about? Can you turn your sense of wonder into a testable hypothesis? What kind of data would you need to validate or refute your hypothesis?

(2) **Identify the relevant variable in the problem.** What quantity are you trying to understand in your hypothesis? Is there a different quantity that serves as a good proxy for the main quantity you are after?

(3) **Identify limitations of the model.** If a model is made from data from lab experiments, for example, will we expect the model to be valid in true living systems?

(4) **Simplify until tractable.** Using what you know biologically, the model might be too complicated to write down. Can you write down a simpler verbal description of what you are trying to model? E.g., can you group several systems in the body into one big class?

(5) **Relate the variables mathematically.** Can you write down the algebraic descriptions of the model? E.g., is the rate proportional to the concentration of the drug?

(6) **Solve or approximate.** Most likely you will use numerical methods such as Euler's method that we saw in Chapter 6 and will see in Chapter 14.

(7) **Does the model provide added value?** What have you learned by modeling this problem with the model you have developed? Often it will belong to one of the three kinds of added value described above but it does not have to.

(8) **Tweak model and compare solutions to each other and to empirical data.** How does the agreement between the model and the data change? Does it improve? How is the agreement on the edge cases (e.g., a small time after or a long time after)?

Of course this list is by no means exhaustive but we feel that it at least gives a model-designer some guidance as to how to get started thinking about their model.

7.3. Approach in this book

Our goal in this text is to think of mathematical models as equations so that we can use the tools of calculus and numerical analysis to make sense of the model and, by extension, the value being added by the model to the problem being considered. This is not the only way to think of such models but it is the one we will be using in this text.

7.3.1. Model formulation. Our main goal when formulating the model is to be able to write down expressions that can then be analyzed using the tools of mathematics. In our context, we are looking for an algebraic description of our model and so we describe things with that in mind.

- **Verbally:** Write down a narrative description of what is going on in the system. The description should be biologically accurate and suggestive of the underlying mathematics going on. See the text in Section 4.2 for an example of such a description.
- **Visually:** All the models in the text can be represented as (perhaps very complicated) box diagrams. Draw the diagram for your model, being sure to label the boxes and arrows between them very carefully. For example, see Figure 3.1.

- **Algebraically:** With the box diagram in hand it will prove pretty straightforward to write down the equations that define the model. Before we write down the equations in the notation that we have chosen, we find it helpful to write down an equation with a verbal description of the quantities involved instead of just using the notation. For example, see (5.2) and (5.1).

7.3.2. Model analysis. Once we have written down a model that can be analyzed mathematically, we ask ourselves a series of standard questions. Again, in our context, we have certain kinds of models in mind. Even so, we still think that the questions are general enough to be widely applicable.

- **Graphically** We use the visual representations discussed earlier in the book to see if the models make sense: phase diagrams, plots of particular solutions, etc. See the various plots in Chapter 5 for a variety of examples.
- **Numerically** We consider the numbers output by the model. Are they of the right sign? Of the right scale? Are they consistent with empirical data in the literature? See, for example, the plots of catfish data and model predictions in Chapter 4.
- **Physically** We consider special behaviors of the system. What is the long-term behavior? What are the equilibria? Are they stable? See, for instance, the way we came up with models that describe the Allee effects in Chapter 5.

7.4. The biological models in this book

Up to now in this text, we have seen an important class of biological problems that can be better understood through the development of an accurate model: the study of population growth. Problems involving growth of populations have been of interest to biologists for a long time: starting in the 18th century with Malthus and his exponential growth model, continuing with Velhurst and his logistic model in the mid-19th century, Lotka and Volterra with their predator prey models (see Chapter 18 for more on this) and MacArthur and Wilson with their island biogeography model. In each case a great deal was learned about the system being studied both in the process of developing these models and in applying them later.

Another important class of biological problems is related to questions about the distribution of medicines and the like in a person's body; as mentioned above, this is the field of pharmacokinetics. This visualizes the human body as a series of compartments with the flow-rates between the compartments. These flow-rates can be measured empirically as can the concentration of the drug in the various compartments. It is possible to discover unexpected ways in which the drug is distributed as well as what mechanisms the body uses to break down the medicine.

A third, and final for this text, class of biological problems we can study in this way is that of the spread of disease: this field is known as epidemiology. Epidemiology can be studied in a variety ways (e.g., empirically, statistically, etc.) but it can also be studied analytically–this is what we do in this text. We develop systems of differential equations based on a box model that predicted the spread of disease and the number of people infected by the disease.

It is perhaps surprising that these seemingly different problems can be modeled in similar ways but since all the problems are related to how certain quantities change with time, it is true that differential equations can be used in all three of these cases. Since these kinds of models are so widespread, we feel justified in this Section having given general advice but keeping box models in mind.

CHAPTER 8

Research Interlude: Reading a Research Paper

Different texts require different kinds of reading. Reading a research paper is very different than most kinds of reading we do day-to-day. Because reading a research paper[1] is so different, we think that it would be helpful to describe *a* general approach to reading one. We start, though, by describing how to even go about finding a research paper to read.

8.1. Finding existing models of something

A modeler of biological systems is necessarily heavily dependent on the research literature in both mathematics and biology. This is a broad literature, with over 300 journals in applied mathematics alone and far more in biology. In a series of chapters we hope to guide you in the process of using the literature to build, parametrize, support, and analyze your model. We will use as an example the development of a paper modeling the life cycle of the mosquito that carries Zika virus.

The first issue that must be addressed is, **"What counts as a reference?"** Probably your first stop in forming an idea for your paper is Wikipedia's entry on Zika. Wikipedia is not a peer reviewed publisher of actual research. It is a summary of what is known on a subject and can be helpful, but it really cannot be used as a point of reference in a research paper. But let's say you read the Wikipedia entry. From this you learn that the most common vector of Zika virus is the *Aedes aegypti* mosquito. The Latin name of this organism is hugely important in searching for scientific research articles about it. Then you look up *Aedes aegypti* in Wikipedia, and possibly "mosquito" as well, and learn the basics of its life cycle.

Your next stop may be **Google scholar** (which some librarians don't care for and would rather you use the library search engines, but that's another story!). You find 121,000 results from a search on the phrase "*Aedes aegypti*". So before you even read a research paper, you have to deal with the question: Which one? You can't read all of them.

So before you even read a paper, you need a question in mind that will point you toward the ones that are useful. If your research project is building a model of the lifecycle of this insect, an appropriate first question is, **"Has anybody else done this already?"** Narrowing the Google scholar search by using the phrase, "aedes aegypti mosquito mathematical model" reduces the number of articles to a mere 8,710 results. But glancing at the first page of hits tells you that people are modeling all kinds of things. You are interested in the life cycle, so you need

---

[1]In this chapter, we are describing the process of finding a research paper which can be understood and analyzed using the methods of this book. There are papers that are purely mathematical or analytical and the advice for those papers would be a little different and the set of papers of that kind that can be read depends heavily on the background of the reader.

to **narrow it down**. The phrase "aedes aegypti mosquito life cycle mathematical model" cuts it down to about 5,000 entries. Some have to do with temperature, some have to do with disease transmission. You want population dynamics, so you put that in also and search on "aedes aegypti mosquito life cycle mathematical model population dynamics" getting about 3,000 hits. **Reading the first few pages of hits**, you see that some models are probabilistic (the word "stochastic" is a tipoff), some are for a different insect (e.g., *Aedes albopictus*), some are more about disease transmission (with the word "dengue" in the title, for example) and some are physiological models. But on the very first page two articles appear:

- Focks, D. Ae, et al. "Dynamic life table model for Aedes aegypti (Diptera: Culicidae): analysis of the literature and model development." *Journal of medical entomology* 30.6 (1993): 1003-1017.
- Dye, Christopher. "Models for the population dynamics of the yellow fever mosquito, Aedes aegypti." *The Journal of Animal Ecology* 53.1 (1984): 247-268.

Furthermore you notice that the Focks *et al.* article has been cited by 353 other articles. It dates back to 1993, so any modelers since then probably mentioned it in their paper. You **check these citations**, which is easy to do with Google scholar. You notice many of them are not mathematical in nature. You click on the **"search within citing articles"** box, and you search on "mathematical model life cycle". There are 124 results. Now the work begins.

Each entry in Google scholar has a few sentences following it. These will help you weed out articles that are not relevant to the question, "Who has modeled the life cycle of this insect?" On the first page of the search are 4 articles that look like they might be relevant. Click on each one and bring up the abstract for the article. **Keep a document going as you review** all 124 results, including the citation information and abstracts of those you find useful. From this list you will eventually select a dozen or so to review. But, as a starting place, you decide to look at the original Fock article. You download it and attempt to figure out what they have done.

First, you **read the abstract**. It is clear that the authors are interested in multiple factors in the life cycle: temperature, habitat and food availability, weather (in particular rainfall). The authors describe their model as a "life table" model. You look up "life table" in Wikipedia and discover it's a table of mortality rates at different ages. So the authors are separating mosquitoes by age and tracking death as the insects pass through their various stages. You want to do this also! But how did they do this? The next step is to **scan the article for actual equations**. To your disappointment, you don't find many, although there is a pair of differential equations describing the change in weight of a larvae in terms of food supply. Although this could be quite useful, it's not exactly what you were planning to do. But how did they actually do the simulations? You next look at the "numeric calculations" section, where it appears they did a combination of differential equations describing larval development in a day, alternating with daily updates on temperature, food supply, etc. All of this was using a piece of software that you don't have. In fact, this article is so old that the computing power available was probably not able to process a full set of ODEs for a long time period, hence the daily updates.

You decide to look for more recent articles that might use techniques similar to those in this text. After restricting your search to "since 2012" you find this:

- Lana, Raquel M. et al. "Seasonal and nonseasonal dynamics of Aedes aegypti in Rio de Janeiro, Brazil: Fitting mathematical models to trap data." Acta tropica 129 (2014): 25-32.

The abstract clearly indicates that the authors are trying to fit a population dynamics model to data. You scan for equations and figures. **One figure is a box model** that includes eggs, larvae, pupae and adults. You find 4 equations, all for $\frac{dE}{dt}$. This is very confusing. Reading the paragraph before these equations makes it clear that they are supposed to be for E, L, P, W (eggs, larvae, pupae, adults) respectively. Staring at them some more, the right-hand side of these equations make sense. So there was a big typo in this paper! Alas, not as unusual as one would wish. Aside from this, the paper is doing something quite similar to what you are interested in doing, and successfully matching some data as well. You decide to make this paper the key to your project. You will let it lead you to the relevant models in the literature, and give you ideas for useful extensions and variations of the model used by those authors.

When you write your paper, you will describe the models of Fock *et al.* and Lana *et al.*, as well as a few related ones, while admitting that your review is not exhaustive. Then you will move on to your own model and how you altered, extended, or found new applications of the one by Lana *et al.*

8.1.1. Summary: To find models others have done

(1) Use Wikipedia to learn the basic facts about and Latin names of relevant organisms.
(2) Use the Latin names and keywords describing your modeling interest in Google scholar.
(3) Narrow your search with more keywords.
(4) Find an early article that looks relevant with lots of citations.
(5) Search among the citations for relevant articles.
(6) Keep a record of these!
(7) Pick something to start with that looks promising.
(8) Read the abstract to see what the authors claim to have done.
(9) Scan the article for equations (if that is what you care about at this point).
(10) If the authors do not make it possible to understand or reproduce their results, try another paper.
(11) Again, skim the paper; read the abstract, look for a box model, look for equations.

8.2. Close reading

Suppose you have now identified a paper that you know is really relevant to what you want to do and which deserves a careful read. How should you go about this? Most authors have made an effort to write linearly, so that one section leads naturally to the next. And yet, that is probably not the best way to approach a paper.

First, **read the abstract**. This is a summary of what the authors feel is the most important take-away of the paper. You might be looking for something different, but knowing what they are trying to prove helps your understanding of the paper.

Second, **look at the figures**. What quantitative information is given in the paper? What things are compared with each other? Knowing this before reading the text is really useful because it allows you to understand the experimental design better as it's being explained.

Now, **read the paper in order**.

The background (or introduction) is very useful, especially if you are a beginner in the field. It may lead you to other papers of interest. It may teach you words that are useful in searching for other papers.

The "methods" or "model development" section explains how the work was organized and done. The mark of good research is "reproducible results". In experimental research that means that the methodology is so well described that you (if you had resources) could duplicate the experiment and presumably get the same results. In modeling, this means that you can program the exact same model as the authors with the same parameters and initial conditions and get the same exact answer. As you read, track whether the authors have given you enough information to do this. Many do not, but if they do you have a valuable starting point for extending or altering their model.

The "results" section describes the outcome of experiments, whether physical or numerical. These are the answers to the questions posed in the introduction. As you read this you might consider what other sorts of "results" could be pursued in your problem.

Finally, the "conclusions" or "discussion" section ought to place the entire study in the broader context of what is known in the field. Here is where you learn what debates and questions are framing the entire research area.

The purpose of a close reading is three-fold. The first goal is to understand the study to the point of reproducibility. The second goal is to gain a better overall understanding of the field, which will surely be new for you. The third goal is to learn to write well. A close reading of a well-written research paper will be a good influence on you when you write your own.

8.2.1. Summary: Doing a close reading

(1) Read the abstract.
(2) Study the figures.
(3) Read the paper in order.

8.3. Finding parameter estimates for your model

One of the most important, and also most frustrating, uses of research literature in modeling is in the parametrization of your own model. We continue the example of modeling population dynamics of *Aedes aegypti* based on the paper:

- Lana, Raquel M. et al. "Seasonal and nonseasonal dynamics of Aedes aegypti in Rio de Janeiro, Brazil: Fitting mathematical models to trap data." Acta tropica 129 (2014): 25-32.

Let us suppose that, after much thoughtful reading of the biology of this organism and the close reading of that paper, you decide that you can justify a model of the form:

$$E' = bW - m_E E - d_E E$$

$$L' = m_E E - m_L L - d_L L - f_L \frac{L^2}{K}$$
$$P' = m_L L - m_P P - d_P P$$
$$A' = m_P P - d_A A$$

You wish to compare this model to one of the four in the Lana *et al.* article. But instead of parametrizing the two by data fitting, you would like to know apriori estimates of all of the parameters in the model. For example, m_L is a maturation rate of larvae. You can see this in the equations above, as this amount leaves the larvae compartment and enters the pupae compartment. But what might this rate be? Let's say your time units are days and your insect units are numbers of individuals. Then the units of m_L must be (percent) per day, so that $m_L L$ becomes (percent of) larvae maturing per day.

After a bit of searching using Wikipedia and Google, you will realize that searching on "maturation time," "maturation rate," "development time," and "development rate" all lead to similar sorts of studies, and you find a few of these that look promising. How can you tell if they will have the information you need? The abstract tells you that they are studying the correct species and looking at maturation. But have they actually taken observations of the time spent as a larva? To find most of the useful numbers needed for modeling, a good strategy is to look at the figures in a paper next. If a research group has measured any interesting quantity, they will almost always display it in a figure. You might find maturation times for different temperatures. You might find maturation times split into four different measurements, as the larvae actually molt four times. These stages, called *instars*, are often counted in field measurements, which you might find as well, usually displayed in a table. So, look at the tables, too.

Once you believe you have found a paper that gives you the information you need, the next step is to read the paper closely so you understand how the data was obtained or measured. If the authors convince you they did a good job, then use their paper as a basis for your parameter. If not, find another paper.

8.3.1. Summary: To identify papers that help you estimate a parameter
(1) Read the abstract.
(2) Look at the figures.
(3) Look at the tables.
(4) If the paper seems to have what you need, read the rest closely.

CHAPTER 9

Brief Introduction to Sage

Sage [2] is a free, open-source computer algebra system built on top of the *very* mainstream programing language Python. It is meant to be an alternative to Mathematica, MATLAB and Maple all of which are non-free, closed-source and languages unto themselves. There is an active Sage community online and there is an excellent book written for undergraduates using Sage [1]. We emphasize the use of Sage in this book as it has the necessary functionality for all the computations you might need to carry out in projects, assignments, and the like.

9.1. Try it out!

Go to

http://sagecell.sagemath.org/

and type

```
plot(3*x^2,(x,0,3))
```

and click the big evaluate button. You should see a graph of the parabola $y = 3x^2$ with vertex at $(0,0)$, for x between 0 and 3. How would you graph $\sin(2x)$ for x between -1 and 3π? How would you change the color, thickness, and style of the line? Try

```
plot(sin(2*x),(x,-1,3*pi),thickness=3,color="red",linestyle="--")
```

For the `linestyle` parameter you have the following choices:

```
linestyle="-"
linestyle="--"
linestyle=":"
linestyle="-."
```

Try them all out for other functions and ranges of x values.

For the `color` parameter, you can enter one of 148 common colors. Here are the first bunch:

```
'aliceblue'
'antiquewhite'
'aqua'
'aquamarine'
'automatic'
'azure'
'beige'
'bisque'
'black'
```

and you can see the entire list in alphabetical order by evaluating

```
sorted(colors)
```

in a cell.

Sage can do more than plotting. Try the following and try to predict what they do:

```
solve([x^2-16==0],x)
```

```
var('x y')
solve([x+y==4,x-y==2],x,y)
```

```
find_root(2*cos(x)-e^x,-1,1)
```

A little explanation of the notation might be useful. In the first one

```
solve([x^2-16==0],x)
```

notice the double equal sign. This is the equal sign of "these two things are equal" which is different than an equal sign like

```
y=3
y^2
```

which means "set the variable y equal to 3". The `solve` command takes in two things: first a list of equations to be solved. Then the variables for which you want to solve. Maybe this is more easily seen in the second example

```
var('x y')
solve([x+y==4,x-y==2],x,y)
```

where we first tell Sage that we have two variables x and y. Then we ask Sage to solve the equations

$$x + y = 4$$
$$x - y = 2$$

for the variables x and y.

How do you think you might ask Sage to solve?

$$x + y + z = 3$$
$$x - y = 2$$
$$y + z = 1$$

Something like

```
var('x y z')
solve([x+y+z==3,x-y==2,y+z==1],x,y,z)
```

will work.

9.2. Getting Sage

You might imagine that the Sage cell server that you've tried will not be so good for larger computations like the ones you might do in the projects below. You have two options.

(1) You can install it on your own computer. Go to

http://www.sagemath.org/download.html

for instructions on how to do this. This can be a little tricky so it is advisable that you do this only if you are pretty comfortable with installing unfamiliar software on computers. It is easiest to do with Mac OS computers and Linux computers. Once you have it installed and figure out how to run Sage from the command line (open a terminal and type "sage"), the easiest way to get started is to type "notebook()" at the `sage:` prompt. This will open a page in your default browser that will have a list of projects (this list will be empty at the beginning). Projects themselves are made up of a bunch of cells like on the Sage cell server.

(2) You can access Sage on the cloud through CoCalc. Go to

$$\text{https://cocalc.com/}$$

and sign up for a free account. Again your work will be divided into projects and each project is made up of a bunch of cells like the Sage cell server.

It can be a little challenging getting Sage up and going on your own but you can find a number of getting started with Sage videos by searching online.

9.3. Solving and plotting differential equations in Sage

Most of this course will be about solving systems of differential equations and Sage is quite good at doing this. Let's look at an example of how to solve a particular system of differential equations in Sage. Here is some sample code

```
N,t = var('N t')
ss = desolve_rk4(0.9*N*(1-N/10),N,ivar=t,ics=[0.1,0.2],end_points=20,
step=1)
list_plot(ss,plotjoined=True,thickness=3)
```

The first thing we do is declare two variables N which will be the population and t time, the independent variable.

The second thing we do is call the differential equation solver. The equation we are solving is the logistic equation

$$\frac{dN}{dt} = 0.9N(1 - N/10).$$

The right hand side of this equation is the first variable that the solver takes in (note: the rk4 in the name of the solver refers to the fact that it uses the Runge-Kutta solution method instead of, say, Euler's method). The second variable it takes in, N, is the dependent variable for which we want to solve. The next variable `ivar` is the independent variable and the one after that is the list of initial conditions where the first number in the list is the initial t value and the second is the value of N at that initial t value. In particular, the list `[0.1,0.2]` encodes the statement $N(0.1) = 0.2$. The last two variables tell the solver at what t value the solver should stop and the step variable says how big an h (in the notation of Chapter 6) to use. Try this in Sage and try a few variants. See that the expected things happen when you change the intrinsic growth rate 0.9 to something else and the carrying capacity of 10 to something else.

The third and final thing we do is plot the results of the solver. The solver returns a list of points $(t_0, N_0), (t_1, N_1), \ldots$ and the function `list_plot` tells Sage to plot the points. To get a curve we set the variable `plotjoined` to be `True`.

9.4. Basic Sage programing

In Sage, you can write your own functions. We implement Euler's method as described in Chapter 6 to illustrate how to do this. Here is a short code snippet with some very non-optimized Sage code for Euler's method:

```
def my_euler(de, ics, end_points, step):
    t = ics[0]
    N = ics[1]
    output = [(t,N)]
    while t < end_points:
        t = t + step
        N = N + step*de(t=t,N=N)
        output.append((t,N))
    return output
```

It can be used, for example, as follows:

```
N,t = var('N t')
g = N*(1-N)
P = my_euler(g,[0.1,0.2],10,0.1)
list_plot(P)
```

Here the differential equation is the logistic equation

$$\frac{dN}{dt} = N(1 - N)$$

and the initial conditions are $N(0.1) = 0.2$, iterations are carried out until $t = 10$ and the step size is $h = 0.1$. Now we dissect the snippet line by line:

```
def my_euler(de, ics, end_points, step):
```

First, do not forget the colon at the end!!! This line is telling sage that you are about to define (`def`) a function called `my_euler` and that it will take in 4 parameters. At this point Sage does not care what they are, just that there are 4. The names of the variables should be descriptive to a human reader: the variable `de` is the right-hand side of the differential equation you are solving; the variable `ics` is a list of the initial conditions (the initial value of the independent variable goes first); the variable `end_points` represents which is the last value independent variable takes on; and the variable `step` is the h in Euler's method as describe in Chapter 6.

Remember that the Euler's method starts with the initial conditions and replaces t with $t + h$ and N with $N + hf(t, N(t))$. The points (t, N) computed this way can be plotted to give a graphical representation of a numerical solution to a differential equation of the form

$$\frac{dN}{dt} = f(t, N).$$

The first three lines

```
t = ics[0]
N = ics[1]
output = [(t,N)]
```

respectively tell Sage that you want the independent variable to be set to its initial value, the dependent variable to be set to its initial value and the list of points (called `output`) that you want the function to calculate be assigned its first element, namely the point corresponding to the initial condition (notice the parentheses around the point).

The next part of the code

```
while t < end_points:
    t = t + step
```

```
N = N + step*de(t=t,N=N)
output.append((t,N))
```

is more interesting. A `while` loop will keep doing what is below it until a condition is met. First, let's describe what is inside the loop: the line `t+step` is telling Sage that, like in Euler's method, the variable t should be updated to $t + h$, N should be updated to $N + h * f(t, N(t))$ (here f is denoted by `de`) and the updated point is added to the list of output points. The question is how long should we repeat this process. We want to keep updating points until the variable t passes the `end_points` parameter. This is what a `while` loop does!

We will revisit coding periodically throughout the book, but we hope this will get you off to a good start. Since Sage is based on Python, by searching for Python tutorials online, you will be able to find lots of advice about how to code in Python (and therefore Sage).

Bibliography

[1] Gregory V Bard. *Sage for Undergraduates*, volume 87. American Mathematical Soc., 2015.

[2] W. A. Stein et al. *Sage Mathematics Software (Version 7.3.1)*. The Sage Development Team, 2017. `http://www.sagemath.org`.

CHAPTER 10

Projects for Population Modeling

Up to now you have seen exercises (straightforward questions to check your understanding of the reading) and problems (more open-ended questions designed to stretch your understanding of that chapter's content. A project is something different. It is meant to have you dig into the scientific literature and ask and answer your own questions based on what you find. The list below is meant to be indicative of the kinds of general questions that can be asked about modeling populations and we encourage you to find your own questions to answer.

(1) The guacamaya (*Rhynchopsitta pachyrhycha*) is a parrot native to the Sierra Madre Occidental in Mexico. It feeds largely on conifer seeds, and because of habitat loss due to logging, these birds have lost much of their original range and have dropped to only about 1500 breeding pairs in a few large colonies in the mountains of Mexico.

   The populations of these birds appear to exhibit the Allee effect. These parrots congregate in large social groups for almost all of their activities to better watch out for predators. When the population drops below a certain number, then these birds become easy targets for predators adversely affecting their ability to sustain a breeding colony.

   A potential model for this population given by [2] is

$$\frac{dN}{dt} = N(r - a(N - b)^2),$$

   for $r = 0.04$, $a = 10^{-8}$ and $b = 2200$ (measure t in years).
   (a) What are the units of these constants? Justify your answer.
   (b) What are the equilibrium populations for this population? What kind of equilibria are they? Use a phase portrait to justify your answer. Plot the phase portrait in Sage.
   (c) Justify both biologically and mathematically that this population exhibits a strong Allee effect.
   (d) Plot a solution to this differential equation using the initial condition $(0, 1500)$ and reach a conclusion about the current population of 1500 parrots.
(2) Study the model in the previous problem a little more generally. In particular, write a paper in which you describe the physical interpretation of the three parameters a, r and b and in which you study the role the three parameters have in the equilibria of the system and the extinction thresholds. After determining which of the three parameters has the biggest effect on the populations dynamics, conduct some research on what interventions might be possible to change the value of the parameter.

(3) A population of sand-hill cranes has been modeled by a logistic equation with constant harvesting

$$\frac{dx}{dt} = rx(1 - x/K) - H$$

with carrying capacity of $K = 194,600$ members and a growth rate of $r = 0.0987$ yr^{-1} [1]. Find the critical harvest rate for which constant harvesting will drive the population to extinction, and find the equilibrium population size under constant harvesting of 3000 birds per year. Use Sage to analyze the sand-hill crane model.

(a) Show what happens to the population if the harvesting rate is above and below the critical rate (give a graph in Sage with two plots on it, one from each case).

(b) Assume that the habitat of the sand-hill cranes is disrupted by development and so the carrying capacity is decreased to half the former value. How does this effect the evolution of the population over time when subjected to the same harvesting that you used in the first part of this problem (here give a graph from Sage with two plots - one for each carrying capacity)?

(4) An improvement to Euler's method described in Chapter 6 is to update the new values of t and N differently. The midpoint method has the update of N occur via the formula

$$N(t + h) = N(t) + hf(t + h/2, N + hf(t, N)/2).$$

Implement this improved version of Euler's method. The definition of your function should be

```
def my_euler_midpoint(de, ics, end_points, step):
```

(5) In Chapter 5 we studied a model for the amount catfish in Lake Victoria and we observed that neither exponential models nor logistic models fit the data well. A feature of the data was that it started off as logistic growth, leveled off (as expected) and then decreased rapidly back towards the horizontal axis. What happened biologically explains this feature: with the introduction of the aggressively predatory Nile perch, the population of catfish bottom out. A way to model the introduction of a known predator at a certain time is something like

$$\frac{dN}{dt} = rN(1 - N/K) - hNu_T(t);$$

this looks like it models a population that grows logistically and is harvested proportionally. The difference between this model and the ones with harvesting in Chapter 5 is this function $u_T(t)$ (which is 0 when $t \leq T$ and becomes 1 when $T = t$ and stays 1 for $t > T$. This has the effect of saying there is no predation before time T and proportional predation after time T. In this case, it is known when the Nile perch was introduced and at what time they were present in large enough numbers to have a noticeable effect on the catfish population. Such a function u is called a step-function.

The catfish data are:

(1965 , 4.24)

```
(1966 , 9.78)
(1967 , 27.60)
(1968 , 15.00)
(1969 , 17 10)
(1970 , 26.70)
(1971 , 29.00)
(1972 , 31.20)
(1973 , 31.90)
(1974 , 35.20)
(1975 , 21.30)
(1976 , 39.50)
(1977 , 31.50)
(1978 , 31.50)
(1979 , 39.00)
(1980 , 38.50)
(1981 , 22.80)
(1982 , 30.00)
(1983 , 12.60)
(1984 , 39.40)
(1985 , 27.50)
(1986 , 16.30)
(1987 , 7.58)
(1988 , 0.19)
(1989 , 0.69)
(1990 , 0.65)
```

In Sage such a differential equation could be solved via something like

```
desolve_rk4(A*N*(1-N/K)
        -B*(unit_step(t-T)-unit_step(t-1990))*N,
            N,ivar=t,ics=[1965,4],end_points=1990,step=1)
```

where the parameter A is the intrinsic growth rate of the catfish, the parameter K is the carrying capacity of the population, the parameter B is the harvesting rate starting in year T which is itself the last parameter in the model. Find reasonable choices for these four parameters and plot your fit and data together to get something like in Figure 10.1.

(6) Imagine a small herd of cows in a field of modest size. The following example shows how the initial condition might affect the final outcome. May developed a model to describe the dynamics of the amount of vegetation V in the field [3]:

$$\frac{dV}{dt} = rV(1 - V/K) - \frac{HbV^2}{V_0^2 + V^2}$$

where $r = 1/3$, $K = 25$, $b = 0.1$, $V_0 = 3$, $H = 10$.

(a) Find all possible equilibria and determine their stability (notice that algebraically you can use the solve function in Sage to do this).

(b) Illustrate these equilibria by solving the differential in Sage with different initial conditions and plotting on the same graph.

(c) How much does this change if $H = 20$? $H = 30$?

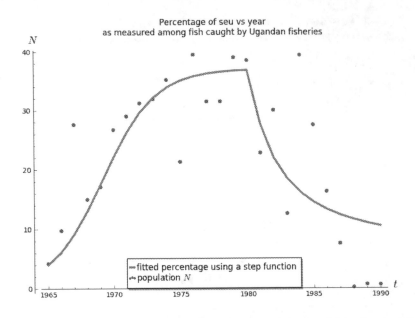

FIGURE 10.1. A potential plot of a solution to the model in Project 5.

    (d) Figure out what the parameters mean biologically. Conduct some
        research to try and figure out how such parameters might be deter-
        mined experimentally. Explain why the V and V_0 terms are squared
        in the harvesting term. What happens if you cube them or have them
        to the first power?

Bibliography

[1] Fred Brauer, Carlos Castillo-Chavez, and Carlos Castillo-Chavez. *Mathematical
models in population biology and epidemiology*, volume 40. Springer, 2001.

[2] Joseph M. Mahaffy. Math 122 - Calculus for Biology II - Fall Semester, 2004 -
Qualitative Analysis of Differential Equations - Examples. `http://jmahaffy.`
`sdsu.edu/courses/f00/math122/lectures/qual_de/qualdeeq.html`. Acces-
sed: 12/27/2016.

[3] Robert McCredie May. *Stability and complexity in model ecosystems*, volume 6.
Princeton University Press, 1973.

Part 3

Drug modeling

CHAPTER 11

Introduction to Pharmacokinetics

When a patient takes a dose of medicine (or when a person inadvertently consumes some amount of toxin), it is important to know the medicine's disposition: where it goes and how quickly it travels there. The human body is a complicated system and there are many parts of the body involved in the absorption, distribution, metabolism and excretion of things ingested by a person. Depending on the compound being administered, only certain parts of the body become saturated with it and so the complicated human body can be simplified to a small number of compartments with arrows between them indicating the flow of the compound from one compartment to the other.

Such a conceptual model combined with data collected in the lab allows one to write down a system of differential equations. With the algebraic representation of the model in hand, a researcher can then make predictions and better understand the mechanisms involved.

A simple model is a model in which the human body is a single compartment and a question to be asked is how quickly will the compound be eliminated from that compartment. The number that describes this amount of time is the *elimination half-life* and is denoted $t^{1/2}$. In this chapter we discuss how to calculate the half-life by using Pentamidine, an anti-microbial drug used in the treatment of African sleeping sickness, as an example.

11.1. Biological context

For almost 20 years, starting in 1986, the citizens of Uganda suffered from ongoing civil war. Numerous factions battled in various regions of the country, with 400,000 people left homeless. Near the start of the unrest, Doctors Without Borders established a project in northern Uganda. The mission of Doctors Without Borders is to provide aid to "people whose survival is threatened by violence, neglect, or catastrophe, primarily due to armed conflict, epidemics, malnutrition, exclusion from health care, or natural disasters." Ugandan citizens clearly experienced more than one of these conditions, many of which are linked explicitly to increased transmission rates for infectious diseases.

In particular, Uganda was ripe for a resurgence of sleeping sickness. According to the World Health Organization [7],

> Rural populations living in regions where transmission occurs
> and which depend on agriculture, fishing, animal husbandry or
> hunting are the most exposed to the bite of the tsetse fly and
> therefore to the disease...Many of the affected populations live

in remote rural areas with limited access to adequate health services, which complicates the surveillance and therefore the diagnosis and treatment of cases. In addition, displacement of populations, war and poverty are important factors that facilitate transmission.

Doctors Without Borders began treating patients with sleeping sickness through its northern Uganda project. Because of the prevalence of the disease, the organization eventually became "responsible for supply and distribution of all sleeping sickness drugs in use worldwide today" [4]. Since the first cases in Uganda, the organization has screened a total of more than 2.4 million people for this disease and has treated over 43,000. In 1998, almost 40,000 cases of sleeping sickness were reported, but estimates were that 300,000 cases were undiagnosed and therefore untreated. In 2009, after continued control efforts, the number of cases reported dropped below 10,000 (9878) for the first time in 50 years. This decline in number of cases has continued with 6314 new cases reported in 2012. However, the estimated number of actual cases is about 20,000 and the estimated population at risk is 65 million people [7]. For a survey of efforts to control sleeping sickness, see [8].

Trypanosomiasis, or sleeping sickness, is a tropical disease because its vector, the tsetse fly, requires warm, humid tropical habitats. The protozoa responsible for the disease undergoes several transformations as it moves from fly to human and back again. The disease cannot be transmitted from human to human, nor from fly to fly. Instead the disease has adapted to take advantage of the fact that the tsetse fly requires regular blood meals to develop a life cycle that moves between hosts.

When humans are not present, other hosts will suffice. In addition to cattle and other domestic animals, game animals such as waterbuck, hertebeest, warthog and impala can also host this parasite, providing an animal reservoir for the disease. Displacement of people into rural areas, long periods of encampment and sleeping outdoors, thus have much potential to increase contact with flies that have been infected from wild or domestic animals.

The first sign of infection with sleeping sickness is a sore at the site of infection, at the fly bite. The sore is a localized infection and does not appear to be as serious as it, in fact, is. Depending on which species of protozoa is causing the lesion, it may remain localized for a couple of weeks to several months, after which it enters the blood stream. At this point the patient will experience periodic fever spikes. These could be mistaken for malaria, which has similar symptoms. The current explanation for these spikes is frequent changes of surface proteins on the protozoa, perhaps the result of mutation within the body. As the human immune system mobilizes in response to one set of proteins and removes protozoa displaying these, another set emerges and grows until the immune response can again recognize and overcome it. This stage of intermittent fever is called the "first stage" of sleeping sickness.

At some point (several weeks to several years depending on which protozoa species is present; there are three) the infection crosses into the central nervous system. This is the "second stage" of infection, characterized by headache, sleep disturbance, and depression, followed by mental deterioration, seizures and palsies, ultimately ending in coma followed by death in 100% of cases. The potentially long first stage of the disease, coupled with its close symptomatic resemblance to malaria, is the reason for screening such large numbers of the population.

There is no preventative medicine for this disease. Treatment depends on which of the two stages the patient has. The first stage is difficult to diagnose but relatively easy to treat with Pentamidine (for one of the species) or Suramin (for the others). Both of these drugs have possible side effects but are much less dangerous than the treatment for second stage trypanosomiasis. Treating the second stage of this disease requires a drug that can cross from the blood stream into the nervous system tissue, as the organism has already done. The most commonly used drug for the second stage is Melarsoprol. An arsenic derivative, this medicine is extremely toxic, producing side effects ranging from myocarditis and renal damage to neural damage and encephalopathy. The World Health Organization estimates that 5% of patients die from the medication while 5% relapse. Doctors Without Borders puts the death rate from Melarsoprol at 5 to 20%. They also remark on how painful the drug is, with patients describing treatment as "burning their veins" [6]. For a general discussion of the treatment of sleeping sickness, see [3].

A doctor who discovers sleeping sickness in its early stage is lucky, because the patient can take Pentamidine (for example) rather than resorting to stronger medicine. This drug is produced by Sanofi-aventis and distributed in collaboration with the World Health Organization. The pharmaceutical firm Sanofi-aventis described itself in 2009 as "the only industrial partner engaged in the fight against this disease".

Development of a drug like Pentamidine requires testing both its efficacy and its pathway through the body. All drugs eventually leave the body, whether through metabolism or elimination via the kidneys or liver. In order to give guidelines for doses, the rates at which metabolism and elimination take place must be both measured and modeled. Medicines that must be absorbed into the blood stream and dispersed into tissue have two more processes that must be understood and modeled. Pentamidine is relatively simple, as it is delivered via intravenous bolus and disperses relatively rapidly compared to the rate at which it is removed from the body. So for this drug, a model can safely ignore the processes of absorption and dispersion. The drug does not appear to be metabolized, but is removed from the body through the liver or the kidneys.

Organs that remove toxic substances from the blood stream do so via semi-permeable membranes. In some cases the membranes have special "doorway" through which a molecule can move in one direction only. In other cases the molecules might just diffuse through the membrane. See Figure 11.1 for a visual representation of diffusion across a membrane.

11.2. The model

The quantity of molecules that is passing through the membrane per time unit depends on the chance of a molecule bumping into that membrane. That chance depends proportionally on the concentration of molecules present in the fluid (in this case the blood). This simple relationship is actually describing a differential equation. If $C(t)$ is the concentration of drug in the blood stream at time t, then the rate at which it is being removed is proportional to $C(t)$. That is,

$$C' = -(\text{chance of molecule bumping into the membrane})$$
$$= -kC.$$

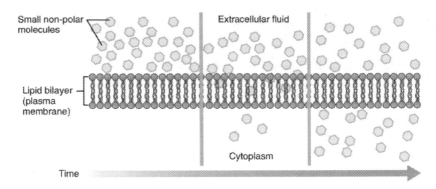

FIGURE 11.1. Representation of diffusion across a semi-permeable membrane. Source: Wikimedia Commons.

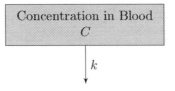

FIGURE 11.2. A one-compartment model of a drug administered intravenously.

In Figure 11.2 we see a visual representation of this model. This model is called a one compartment model and the single compartment represents the concentration in the body's circulatory system. This is the concentration that we can most readily measure and so it is the one that we calculate and model. We observe that there is no drug flowing into the compartment because the intravenous infusion in this model is a one time event. If there was a constant rate of drug flowing into the patient's circulatory system (e.g., if the patient was constantly on an IV or if the medication was taken orally and slowly absorbed from the gut into the circulatory system), then there would be an arrow pointing into the compartment. In this section we are content to consider the simplest of models. The one-time IV dose will be related to the initial concentration of the drug when we set up the initial value problem.

The negative sign indicates that the change is negative and $k > 0$ is some constant that depends on the particular drug and, to a lesser extent, the particular patient. Of course, this is the same equation we studied when we considered population growth, except that the constant, $-k$, is a negative number. Nonetheless calculus tells us the solution:

$$(11.1) \qquad C(t) = (\text{apparent initial concentration}) \times e^{-kt} = C_0 e^{-kt}.$$

Here C_0 is the initial concentration of the drug. In Figure 11.3, we see a sample solution to this differential equation.

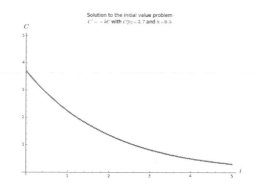

FIGURE 11.3. A solution to the differential equation $C' = -kC$, with $k > 0$. Such a function is said to exhibit exponential decay.

11.3. Half-life

Immediately after receiving the IV bolus, the drug is being removed from the body. At some point there is only half as much left. When is this? We can set up a simple equation:

Half as much left $=$ the value of the solution at some time

$$C_0/2 = C_0 e^{-kT}$$
$$\Rightarrow \frac{1}{2} = e^{-kT}$$
$$\Rightarrow \ln(1/2) = -kT$$
$$\Rightarrow \ln(2) = kT$$

Thus, either $T = \ln(2)/k$ or, equivalently, $k = \ln(2)/T$.

Notice a couple of things. First, T, called the "half-life" of the drug, does not depend on initial concentration, only on k. In the pharmaceutical literature you might not find k, but you might find T instead (this half-life T will often be denoted $t^{1/2}$).

EXERCISE 11.1. What are the units of the $\ln(2)$ in the equations above?

As k, the "constant of elimination", increases, the half-life decreases. This makes sense. If you increase the rate of removal, the drug should spend less time in the blood stream and drop faster. Impaired kidneys could reduce k thus increasing the half-life of the drug. If you are a doctor giving medicine to a person with only one kidney, and if the kidney is the only organ of removal for that medicine, then that person has only half the constant of elimination of a normal person. If the medicine is an IV bolus with simple elimination dynamics like this one, then you had better double the half-life stated in the literature, otherwise you might poison your patient.

What about the initial dose? As C is given as a concentration of drug in the blood, there is some need for translation. Pharmacokinetics texts are careful to point out that the blood is just a proxy for the amount of drug in the entire body. Some of it is in soft tissue and unavailable to be measured. So there is a fudge factor that is merely a fraction: namely the fraction of drug visible to someone taking a blood sample. This can be measured directly in the lab. In other words, if you put 4 mg of drug in an IV bolus and administer it to the patient, wait a bit, then take a blood sample, it may appear that too little is present. In fact, it may appear (based on the model) that the patient has way too much blood, and the drug is quite dilute. For this reason, the initial concentration C_0 in our equation above might really be somewhat less than 4. We say it is C_0/V, where V is the *apparent volume of blood*. V varies with the drug under consideration.

According to the World Health Organization, the recommended dose of pentamidine is 4 mg per kg of body weight. In order to get a certain initial concentration of drug in the body, more must be given to larger people. So a patient weighing 100 kg will receive a dose of 400 mg in an IV bolus. This is to be given once per day. The half-life of pentamidine is 6.4-9.4 hours. This is quite a spread. The apparent volume of pentamidine for IV bolus administration is 140 liters. We will use this number, although it was measured in studies on AIDS patients who may not truly represent a population of trypanosomiasis victims.

So for pentamidine, the equation describing blood concentration over time is this:

$$C' = -kC.$$

Its solution for this dose looks like this:

$$C(t) = (400\text{mg}/140 \text{ liters})e^{(-(\ln(2)/6.4)t}.$$

EXERCISE 11.2. What is the largest possible k according to the WHO and what is the smallest?

11.4. Implications of the model

With this model, we would be able to answer practical questions about dosing schemes. For example, we could answer:

(1) How much pentamidine is left in the body after 24 hours after a single dose?
(2) If the dose is repeated after 24 hours, how much will the initial concentration really be? (taking into account what is left in the body)
(3) If the second dose is given 2 hours early or 2 hours late, how will this affect the initial concentration for that dose?
(4) If the half-life is really 9.4 hours instead of 6.4 (since there seems to be a discrepancy), how do the initial concentrations vary for the second dose?

High concentrations of pentamidine can cause kidney problems. What we really don't want to see in these repeated doses is a constant increase of initial values, driving the concentration higher and higher over an extended time period. How many doses do we have to give before the initial concentration becomes 10 percent greater than what is intended (which can be inferred from the World Health Organization instructions)?

These and other questions are all part of the cure for trypanosomiasis. Recent epidemics in various parts of Africa have reported a prevalence of 50%, surpassing HIV/AIDS in mortality rates in those regions. Pharmaceutical research and the models that inform it are a necessary part of the cure for sleeping sickness, which also encompasses distribution strategies for medical care, the movement of populations, the biology of the tsetse fly and the ecology that supports it. The discussion of pharmacokinetics in this text is similar to material found in any standard pharmacology textbook.

The reader is invited to consult [1] and [2] for further information. Any reader interested in investigating the pharmacology of treatments for tropical diseases such as those discussed in this text can find the relevant constants and treatment regimens in [5].

11.5. One compartment IV administration: Pharmacokinetic details

In most pharmacokinetic models, the drug being modeled is usually eliminated exponentially. A natural question to ask, then, is how do we model drug absorption? This depends on how the drug is administered and what is known about how the drug diffuses throughout the body. The simplest situation to model is one in which the drug is administered intraveneously–this is the simplest situation because we do not have to take into account any absorption by the gut and because a drug's efficacy is determined by its concentration in a patient's blood. This is the model used above to model pentamidine.

In this section, we make three assumptions:

(1) the patient's body can be modeled as a single compartment: that is, the concentration of the drug in extravascular tissues is proportional to its concentration in the blood;
(2) the drug and blood mix instantaneously; and
(3) the elimination follows first order kinetics; i.e., the elimination is modeled by exponential decay.

With these assumptions, we can write down the following differential equation for concentration $C(t)$ at time t:

(11.2) Rate of change of concentration at time $t = C'$

$$= \text{proportional to concentration}$$
$$= -kC$$

where k is some positive constant. Often this constant is denoted k_e where the e stands for "elimination".

So, from what we have seen earlier in this section, we conclude that

(11.3) $$C(t) = C_0 e^{-k_e t},$$

where C_0 is the initial concentration of the drug in the patient's blood.

How is C_0 determined? If we know an initial plasma concentration C_0, using the constant k, we can find $C(t)$ for any time t. Usually, we do not know C_0 but we know an initial dose. In order to translate an initial dose into an initial concentration, we need the volume into which the dose will be distributed. That is, the *apparent volume* of the body. We define this volume as

$$V = \frac{\text{amount of drug in body}}{\text{concentration measured in plasma}}$$
$$= \frac{X(t)}{C(t)}$$

where $X(t)$ is the amount (mass) of drug in the patient's body at time t. The volume V is assumed to be constant (not only does the drug's apparent volume not change with time, for a particular drug the apparent volume is constant for all patients). This will allow us to rewrite model (11.3) without mention of initial concentration:

$$C(t) = C_0 e^{-k_e t}$$
$$= \frac{\text{initial mass}}{V} e^{-k_e t}$$
$$= \frac{\text{dose}}{V} e^{-k_e t}.$$

At this point we have related plasma concentration to dose. To completely understand and apply this model, we need to discuss how to find V. This value is often provided by the drug's manufacturer and is found as follows (similarly, the value k_e is also provided by the manufacturer and will also be found below).

Before we get to the derivation, we look at

$$C(t) = C_0 e^{-k_e t}$$

a little more closely. We take the natural log on both sides and use its properties:

$$C(t) = C_0 e^{-k_e t}$$
$$\Rightarrow \ln C(t) = \ln \left(C_0 e^{-k_e t} \right)$$
$$\Rightarrow \ln C(t) = \ln(C_0) + \ln(e^{-k_e t})$$
$$\Rightarrow \ln C(t) = \ln(C_0) - k_e t.$$

The conclusion to be reached from this derivation is that if we take the natural log of concentrations for a drug that exhibits exponential decay, we should find a *linear* relationship between the natural log concentrations versus time. We will use this fact in the next argument.

Phase I drug trials, have the determination of the drug's pharmacokinetic properties as one of their main goals. The constants k_e and V are determined as follows. Patients are given a drug that is assumed to follow first order kinetics and the drug concentrations are take periodically. See Figure 11.4. Now, we take the natural log of these data and produce a graph that looks linear–see Figure 11.5. The linear regression of these data will produce a line whose slope is k_e and whose y-intercept is $\ln(C_0)$. Once these values are known, then we can write down a model for the concentration versus time only in terms of dose. In Sage, you would write something like

```
ll = [[1, 3.20119738166216], #drug concentrations at times
  [2, 3.00119738166216],        #1, 2, 4, 6 and 10
  [4, 2.60119738166216],
  [6, 2.20119738166216],
  [10, 1.40119738166216]]
var('C,ke,t')
model(t) = C-ke*t
find_fit(ll, model)
```

and it would return

```
[C == 3.401197381662155, ke == 0.19999999999999996]
#to find C_0 we would calculate e^(3.4012)
```

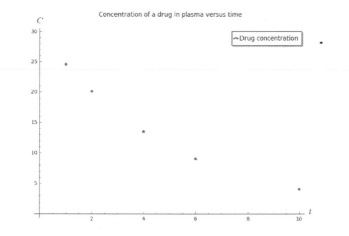

FIGURE 11.4. Plasma concentration data for a hypothetical drug.

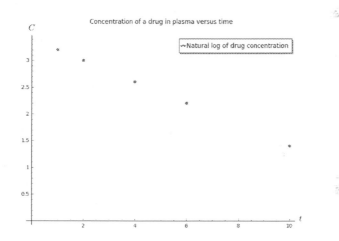

FIGURE 11.5. Natural log of plasma concentration data for a hypothetical drug.

EXERCISE 11.3. Find the concentration 4 hours after 500 mg of a dose with an elimination rate of 0.2 hr^{-1} that distributes into a volume of 30 L.

EXERCISE 11.4. Use the above method to find k_e and V if you know that a 400 mg IV bolus was administered and that the plasma concentration after two hours was 4.5 mg/L and after six hours was 3.7 mg/L. Also, given $k = 0.17$ hr^{-1} and $V = 25$ L, find the dose that should be given to achieve a plasma concentration of 2.4 mg/L at 6 hours.

11.6. Problems

PROBLEM 11.5. Answer the questions about pentamidine asked above in Section 11.4.

PROBLEM 11.6. After an IV bolus of 500 mg, the following data were collected:

t (hr)	1	2	3	4	6	8	10
C (μg/mL)	72	51	33	20	14	9	4

Describe a method to compute k, V and $C(12)$ using only these data.

Bibliography

[1] Arthur J Atkinson Jr, Shiew-Mei Huang, Juan JL Lertora, and Sanford P Markey. *Principles of clinical pharmacology*. Academic Press, 2012.

[2] David W. A. Bourne. Biopharmaceutics and pharmacokinetics. http://www.boomer.org/c/p4/, 2015.

[3] Reto Brun, Johannes Blum, Francois Chappuis, and Christian Burri. Human african trypanosomiasis. *The Lancet*, 375(9709):148–159, 2010.

[4] Medecins Sans Frontieres. Sleeping sickness. `http://www.msf.ca/en/sleeping-sickness`, 2015.

[5] Urban Hellgren, Orjan Ericsson, Yakoub AdenAbdi, and Lars L Gustafsson. *Handbook of drugs for tropical parasitic infections*. CRC Press, 1995.

[6] Integrated Regional Information Networks. DRC: forgotten killer is back. `http://www.irinnews.org/printreport.aspx?reportid=59692`, 2006.

[7] World Health Organization et al. Trypanosomiasis, human African (sleeping sickness), factsheet 259. *World Health Organization, Geneva, Switzerland. http://www.who.int/mediacentre/factsheets/fs259/en*, 2013.

[8] SC Welburn, I Maudlin, and PP Simarro. Controlling sleeping sickness–a review. *Parasitology*, 136(14):1943–1949, 2009.

CHAPTER 12

Two Models for Lead in the Body

A perusal of the publications of the World Health Organization reveals the effects of poverty and ignorance on the health of a region. Ignorance of how a disease works can contribute to its spread. Poverty adds an economic factor to medicine because methods of detection, preventions and cures must be inexpensive to administer on a large scale. The work of the World Health Organization and Doctors Without Borders often must include establishing an infrastructure for delivering care.

Rural areas have special problems because of their relative spatial isolation and endemic disease and insect vectors. People living in rural areas also face economic challenges because of increasingly limited resources for agriculture, the basis of their livelihood. So, a combination of factors encourages people to migrate to cities, where they perceive their chances to be better. Increased population goes hand-in-hand with industrialization: a large concentration of people provides the workforce needed to establish factories as well as the pressing need for employment. Factories and other enterprises with jobs to offer draw people in from the countryside. This process is at work in Kenya, which now has a variety of industries from food processing to the manufacture of glass and chemicals and the assembly of vehicles. Slowly, industry and manufacture are making up an increasing portion of the gross domestic product. Nairobi now has its own stock exchange, listing (as of this writing) more than 60 industries ranging from mining to sugar refineries.

As wealth builds and goods accumulate, a region may experience new environmental hazards or diseases. In fact, we might well find regions experiencing side effects of industrialization that we recognize from earlier experiences in countries such as the U.S. The industrializing of Africa gives human beings a chance to repeat their mistakes. For example, even though leaded gasoline for cars has not been used since the mid-1970s in the U.S. because of the known effects of airborne lead pollution, only recently have African countries begun to go lead-free. The decision to abolish lead from fuel is not completely a case of a belated recognition of known dangers. Rather, it has been a reaction to the development of the effects of widespread exposure: high levels of lead in the blood of many citizens and the appearance of symptoms, especially among children.

12.1. Medical context

Urban children in Nigeria, South Africa, and other African nations showed elevated levels of lead in the blood during studies in the late 1990s. Researchers argued that all of Africa needed to pay attention to this problem, pointing out that a major source of lead is fuels that, in burning, release lead into the air. Air circulates, carrying lead with it to all parts of the continent regardless of local restrictions and laws. The lead passes through lung membranes directly into the plasma, with children having a higher rate of uptake. By 2004 50% of all gasoline in sub-Saharan Africa was unleaded, and Kenya announced it would phase out leaded gasoline by 2006. As an exporter of fuel to the rest of Africa, Kenya's decision would improve the situation throughout the continent.

Leaded fuel is not the only source of airborne lead, however. Lead in paints and other products usually leach into the soil, but can become airborne if these products are burned. Garbage incineration is one way to handle the large amount of waste generated by a city such as Nairobi.

Nairobi's 3 million plus inhabitants, like those of most large cities, produce a prodigious amount of garbage and it has to go somewhere. About 8 miles away is the town of Dandora, which features a 30 acre quarry that Nairobi has been using for a dump, depositing more than 2,000 metric tons of garbage each day until the former quarry is no longer a hole, but a mountain. Any kind of garbage is welcome at Dandora and to manage the incoming deluge, the garbage is burned.

Dandora is a poor area and the dump is a source of potential profit. People, including children, scour the dump for food, reusable materials, and anything they might be able to sell. As they search the dump, these people breathe the fumes from the burning garbage. So in 2007 the United Nations Environment Program, which is headquartered in Nairobi, commissioned a study [4] of the children of Dandora. The study tested 328 children 2-18 years of age who lived near the dump. Over half of them had blood levels of lead in excess of 10 micrograms per deciliter, the internationally accepted "action" level for this poison. Half also had low hemoglobin levels and 30% were anemic, two symptoms of lead poisoning. Lead poisoning also eventually affects the nervous system and brain.

The study also tested soil samples in the area. Of the samples collected, 42% had lead levels that were 10 times higher than what is considered normal. In addition, the soil was contaminated by very high levels of mercury and cadmium. The Nairobi river runs by the dump, so any soil contaminants will eventually leach into it. Water downstream is for crops and residential use. The study was careful to point out that Dandora is not unique. Many cities burn garbage.

Lead poisoning does not produce noticeable symptoms until the blood concentration is fairly high, at which point the victim may experience nausea, fatigue, headache and general weakness. Children may develop learning disabilities, be lower in IQ and smaller in size as a result of prolonged exposure. Advanced neurological symptoms include seizure and coma [6].

There is no easy cure for lead poisoning. Even if a person is removed from all sources of lead, the level of lead in the blood will remain high for a prolonged period. To understand why this is the case we must look at the pharmacokinetics associated to a "dose" of lead. The way toxic substances pass through the body is sometimes called "toxicokinetics", although the approach is identical to the study of any drug.

12.2. The first model

There are two ways in which airborne lead poisoning differs from the pharmacokinetics model we studied in Chapter 11. One is that the lead arrives as a steady ongoing input rather than as an initial dose. The second difference is that the body stores lead in the bones. So there are two important compartments to consider: the blood plasma and the bone tissue. Lead enters the blood through capillaries in the

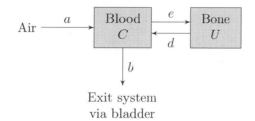

FIGURE 12.1. A first model for lead in the human body.

lungs by diffusion across a membrane, is exchanged between the plasma and bone tissue in both directions, and exits the system from the plasma via the bladder. It is also stored in soft tissue, but for a relatively short amount of time. We will ignore this compartment for now but some models include it.

As discussed before, when faced with a model, one of the first things we should do is fix some notation. Let C be the concentration of lead in the blood plasma, and let U be the concentration of lead in the bone tissue. In the *verbal* description of the model above, we see that we need a constant of absorption a related to how efficiently airborne lead passes through the membranes in the lungs and into the blood, we see that we need a constant d related to how lead passes from blood to bone, a constant e related to how lead is re-released back into the blood from the bone and, finally, a constant b related to how lead exits the system through the bladder. See Figure 12.1 for a *visual* description of the model.

The box model suggests we might need three differential equations for each of air, blood and bone. However, the amount of lead in the air is relatively constant and all we really care about is the arrow representing uptake of lead from the air. So we can leave out the compartment consisting of air. The differential equation for lead concentration in the blood should have four terms according to this picture, corresponding to the four arrows leaving or entering the box. Similarly, the equation for the amount of lead in bone should have two terms.

Movement of lead between compartments has the same basic kinetics as across semipermeable membranes. Rates are just proportional to concentrations or quantities inside the source compartment. However, the units used to measure lead concentration in blood are different from the units used to measure amount of lead stored in the bone. Thus the constants representing transfer of lead from blood to bone are different in the two equations. That is, the constants d_1, d_2 and e_1, e_2 represent the same transfer of lead, but in different units in each of the compartments. So we get the following basic system of equations, our *algebraic* description of the model:

(12.1)

$C' =$ change of plasma concentration

$\quad =$ (environmental exposure) $-$ (urine loss) $-$ (loss to bone) $+$ (gain from bone)

$\quad = a - bC - d_1C + e_1U;$

$U' =$ change of lead in bone

$\quad =$ (gain from plasma) $-$ (loss to plasma)

$$= d_2 C - e_2 U$$

Now we *make sense* of the model in the usual way. First, we try to understand some of the model's *long term behavior* by considering its equilibria. If the bone could not store any lead, we would have only the simple equation:

$$C' = a - bC.$$

No matter what the starting level of lead in the blood, after a while C stabilizes at an equilibrium value. At equilibrium, the change C' is zero, which (using the previous equation) means that $C = a/b$.

If we include the role of the bone, we still get an equilibrium where both C' and U' are zero. Solving the pair of equations

(12.2)
$$0 = a - bC - d_1 C + e_1 U$$
$$0 = d_2 C - e_2 U$$

for the unknown quantities C and U gives us a solution that tells an important part of the story of lead poisoning. The bones pick up lead rapidly and lose it slowly, especially in children. That is, the constant d is large relative to the constant e. In fact, the equilibrium constant of lead in the bones is so large that you would be unlikely to reach it in a lifetime, even with persistent exposure. So although the equilibrium plasma concentration is reached fairly quickly, the lead in the bones continues to increase.

EXERCISE 12.1. Solve system 12.2 to find the equilibria for the model. Compare what you find algebraically to what is asserted in the preceding paragraph.

Second, we discuss finding a numerical solution to the system of differential equations as this will allow us to understand the more immediate behavior of lead in the blood and bone. In order to do this we first need to find values for the constants in (12.1).

12.2.1. Numerical solution to toxicokinetic model of lead. In [2] we find a model for lead, though this is in terms of *amount* of lead instead of *concentration* of lead. Nevertheless we get an analogous system of differential equations except we denote the amount of drug in the bloodstream by X and the amount in the bone by Z. Thus we have the system:

(12.3)
$$\frac{dX}{dt} = -(0.0373 + 0.0039)X + 0.000035Z + 49.3$$
$$\frac{dZ}{dt} = 0.0039X - 0.000035Z$$

with initial conditions $X(0) = Z(0) = 0$. The numbers in this system correspond to the constants in Figure 12.1. In particular,

$$a = 49.3 \qquad b = 0.0373$$
$$e = 0.0038 \quad d = 0.00035.$$

The units of mass are μg and the units of time are days.

With these parameters, using Sage, we get the plot in Figure 12.2.

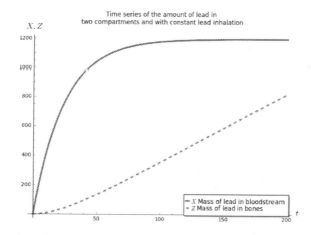

FIGURE 12.2. Plot of a numerical solution of System (12.3) over 100 days. Observe that the mass of lead in the bloodstream already appears to have achieved an equilibrium.

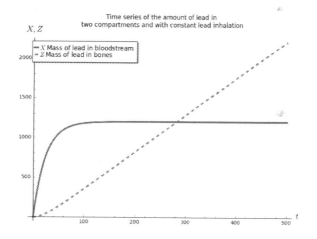

FIGURE 12.3. Plot of a numerical solution of System (12.3) over 500 days. Observe that the mass of lead in the bones continues to grow.

12.2.1.1. *Plotting solutions to systems of differential equations in Sage.* To make Figure 12.2, we used the following code:

```
X,Z,t=var('X Z t')

ss = desolve_system_rk4([-(0.0373+0.0039)*X+0.000035*Z
    +49.3,0.0039*X-0.000035*Z],
       [X,Z],ics=[0,0,0],ivar=t,end_points=200,step=1)

XX=[ [i,j] for i,j,k in ss]
LP=list_plot(XX,plotjoined=True,thickness=3,
  legend_label="$X$ Mass of lead in bloodstream")
```

```
ZZ=[[i,k] for i,j,k in ss]
LP += list_plot(ZZ,color="green",plotjoined=True,thickness=3,
   linestyle="--",legend_label="$Z$ Mass of lead in bones",
   title='Time series of the amount of lead in\n two compartments and
   with constant lead inhalation', axes_labels=['$t$','$X,Z$'])
```

```
show(LP)
```

We explain the code here. The first line

```
X,Z,t=var('X Z t')
```

is declaring three symbolic variables to be used in the code that solves the system of DEs. The first two correspond to the dependent variables (in this case amount of lead in the bloodstream and amount of lead in bones) and the third one to the dependent variable (in this case time).

The second line

```
ss = desolve_system_rk4([-(0.0373+0.0039)*X+0.000035*Z
   +49.3,0.0039*X-0.000035*Z],
      [X,Z],ics=[0,0,0],ivar=t,end_points=200,step=1)
```

is where the system is solved. The function `desolve_system_rk4` uses the Runge-Kutta 4 method for solving a system. The function, as called above, takes in 6 parameters. The first one is a list of differential equations and the second is a list of dependent variables. The list of differential equations is in the same order as the second list: in particular, the first differential equation corresponds to $\frac{dX}{dt}$ and the second to $\frac{dZ}{dt}$ since the second list is `[X,Z]`. The third parameter is the list of initial conditions: the first entry in the list is the initial time, the second is the value of the first dependent variable at that initial time and the third value is the value of the second dependent variable at that initial time. In this case, then, we are saying that $X(0) = 0$ and $Z(0) = 0$. The next parameter indicates which parameter is the independent variable. The last two parameters indicate which is the last t value that will be plotted and the last parameter is the step size.

Based on all these parameters, the function returns a list whose length is `end_points` divided by `step`. Each element of the list is (in this case) a triple of numbers $(t, X(t), Z(t))$, where the last two numbers are approximations of X and Z at time t (which is some multiple of `step`). In order to plot the functions X and Z separately we write

```
XX=[ [i,j] for i,j,k in ss]
LP=list_plot(XX,plotjoined=True,thickness=3,
   legend_label="$X$ Mass of lead in bloodstream")
```

The first line just tells Sage to make a new list of pairs from the list of triples that the DE solver computed. This list `XX` will consist of the points $(t, X(t))$ since of the triples `i,j,k` the solver calculated, we are only collecting the first two numbers in each triple. Then, the second line just starts to build the plot by plotting the points $(t, X(t))$.

The next two lines

```
ZZ=[[i,k] for i,j,k in ss]
LP += list_plot(ZZ,color="green",plotjoined=True,thickness=3,
   linestyle="--",legend_label="$Z$ Mass of lead in bones",
```

```
title='Time series of the amount of lead in\n two compartments and
    with constant lead inhalation', axes_labels=['$t$','$X,Z$'])
```

do the analogous thing to collect and plot the points $(t, Z(t))$. The += tells Sage to place the plot of $(t, Z(t))$ on the same axes on which we had already placed the points $(t, X(t))$.

EXERCISE 12.2. What happens when the source of lead is removed? That means that the constant a is now zero (instead of 49.3). Eventually X will also be zero, but how long will it take? This depends on how much lead has been stored in the bones. If we take values of Figure 12.3 ($X = 1200$, $Z = 2200$), and use these as starting values in the same system with $a = 0$, what happens and how long does it take?

12.3. The second model

The medical intervention for extreme lead poisoning (and other heavy metal poisoning) is called "chelation" and involves chemically removing lead from the blood.

EXERCISE 12.3. Draw a visual representation of this model and compare it to the algebraic representation in the next paragraph.

The patient undergoes treatment for 10-12 weeks during which an IV drip administers chemicals that bind to lead in the blood, increasing the rate at which the kidneys are able to remove it. There are unpleasant side effects to the treatment, including a burning sensation when injected into a vein, fever and chills, headaches, nausea and vomiting, diarrhea, convulsions or seizures, fall in blood pressure, and breathlessness or tightness in the chest. The rate at which the removal takes place just depends linearly on the lead level in the blood. So, in the absence of further exposure (12.1) becomes

(12.4) C' = change of plasma concentration

$\quad\quad$ = (environmental exposure) $-$ (loss due to chelation) $-$ (urine loss) $-$

$\quad\quad\quad$ (loss to bone) + (gain from bone)

$\quad\quad$ = $a - fC - bC - d_1 C + e_1 U;$

$\quad U'$ = change of lead in bone

$\quad\quad$ = (gain from plasma) $-$ (loss to plasma)

$\quad\quad$ = $d_2 C - e_2 U,$

where $f > 0$ is a constant related to how quickly chelation removes lead from the bloodstream.

EXERCISE 12.4. Why is there a negative sign in front of the constant f in (12.4)?

Again, we *make sense* of the model in the usual way.

EXERCISE 12.5. Find the equilibria of the system after chelation has been introduced.

EXERCISE 12.6. Make a plot of this model in which you use a chelation rate $f = 0.2$ in an environment where there is no lead inhalation. Describe how the elimination of lead from the blood with chelation compares to the elimination without chelation.

12.4. Implications

Airborne lead is not the only source of environmental lead in urban Africa. Soil levels and water levels are sources of exposure, as well as occupational exposure. Some diseases, such as HIV, increase susceptibility to lead poisoning. In addition, other heavy metals such as mercury and cadmium create hazards. All of these environmental toxins are related to the industrialization that promises so much for Africa's economic growth, at the same time creating potential health hazards for the citizens who desire that growth.

Lead accumulation is not only a problem in the Lake Victoria region, but has presented challenges worldwide. Mathematical modeling of lead's interaction with the human body is the same everywhere, so though lead studies will account for different inputs of lead worldwide, they still present useful information about the process. For more information on lead in papers with data, information and modeling, see [1, 8, 5].

Lead is not the only heavy metal that poses health risks to life. Mercury, also presents similar problems. The bioaccumulation of mercury in aquatic organisms has been studied extensively. The Daphnia, being low on the food chain presents a starting point for mercury accumulation up the food chain. For information about mercury and Daphnia with modeling see [9, 10, 11].

Because bioaccumulation is prevalent also outside of the Lake Victoria region, studies of other lakes, including Lake Erie, and Lake Murray in Papua New Guinea present useful information. See [3].

12.5. Problems

PROBLEM 12.7. The literature gives constants for transfer rates of lead between blood and bone. See [8, 1] for details. Build a model with these constants and run it numerically on the computer. The accepted "action level" for lead poisoning 10 micrograms per deciliter of plasma. What does the exposure rate have to be to arrive at this equilibrium concentration?

Two children in the Dandora sample had blood concentrations of around 30 micrograms per deciliter. If these children are removed from this environment and placed in an environment with "normal" exposure rate, how long will it be before the level of lead in their plasma falls below 10 micrograms per deciliter?

PROBLEM 12.8. We can extend the model to include transfer of lead between plasma and soft tissue. The literature has constants and models for this situation also. Build a model with this compartment included. Does it differ substantially in its response to the exposure level and recovery times you obtained for Problem 12.7? See [7] for a clear presentation of these constants.

PROBLEM 12.9. Typically chelation is administered for a period of several hours at intervals throughout treatment, which is different from the outcome pictured here. After a single treatment the plasma lead level is seen to drop and then rebound due to release of more lead from the bone. Do the models in Problems 12.7 and

12.8 predict this? How does the time frame of your models compare to what is seen experimentally?

Bibliography

[1] E Batschelet, L Brand, and A Steiner. On the kinetics of lead in the human body. *Journal of Mathematical Biology*, 8(1):15–23, 1979.

[2] Robert L Borelli and Courtney S Coleman. *Differential equations: A modelling perspective*. John Wiley & Sons, 2004.

[3] Karl C Bowles, Simon C Apte, William A Maher, Matthew Kawei, and Ross Smith. Bioaccumulation and biomagnification of mercury in Lake Murray, Papua New Guinea. *Canadian Journal of Fisheries and Aquatic Sciences*, 58 (5):888–897, 2001.

[4] Njoroge Kimani. Environmental Pollution and Impacts on Public Health; Implications of the Dandora Municipal Dumping Site in Nairobi, Kenya. *UNEP, Kenya*, 2007.

[5] Richard W Leggett. An age-specific kinetic model of lead metabolism in humans. *Environmental Health Perspectives*, 101(7):598, 1993.

[6] Herbert L Needleman, Alan Schell, David Bellinger, Alan Leviton, and Elizabeth N Allred. The long-term effects of exposure to low doses of lead in childhood: an 11-year follow-up report. *New England Journal of Medicine*, 322(2):83–88, 1990.

[7] Regents of the University of Michigan. Math 216 Demonstrations - Simplified (Two-compartment) Lead Tracking Model. `https://instruct.math.lsa.umich.edu/lecturedemos/ma216/docs/5_2/`. Accessed: 12/26/2016.

[8] Michael B Rabinowitz, George W Wetherill, and Joel D Kopple. Lead metabolism in the normal human: Stable isotope studies. *Science*, 182(4113):725–727, 1973.

[9] Martin TK Tsui and Wen-Xiong Wang. Maternal transfer efficiency and transgenerational toxicity of methylmercury in Daphnia magna. *Environmental Toxicology and Chemistry*, 23(6):1504–1511, 2004.

[10] Martin TK Tsui and Wen-Xiong Wang. Influences of maternal exposure on the tolerance and physiological performance of Daphnia magna under mercury stress. *Environmental Toxicology and Chemistry*, 24(5):1228–1234, 2005.

[11] Martin Tsz-Ki Tsui and Wen-Xiong Wang. Biokinetics and tolerance development of toxic metals in Daphnia magna. *Environmental Toxicology and Chemistry*, 26(5):1023–1032, 2007.

CHAPTER 13

Methods of Drug Administration

Early European explorers may have relied on Arabic maps and advice in their travels, but they didn't have the advantages of information that we now enjoy. Travel guides, tourist boards, and government websites all offer a wealth of both friendly advice and warnings to those contemplating a visit to Africa. One consistent theme of these warnings is the possibility of getting sick. Visitors to a new place always run the risk of getting sick, just because the variants of local colds, flus, and dysentery are different from those at home and so the visitor has yet to acquire immunity to these strains.

People from prosperous countries away from the tropics live in communities where childhood vaccinations, availability of antibiotics, and a general lack of communicable diseases make them more concerned about those ailments associated with genetics, lifestyle, and advancing age. It is hard for people from North America and Europe to imagine the incidence and variety of old-fashioned infections that occur in tropical regions.

The World Health Organization lists an alarming number of diseases to think about when visiting Africa. In addition to familiar diseases such as malaria, HIV, and cholera, there are many less familiar ailments such as sleeping sickness (trypanosomiasis), dengue fever, ebola, Marburg haemorrhagic fever, schistosomiasis, Rift Valley fever, and yaws. Some diseases are transmitted by insects, some by human contact. Vaccines or other preventative measures exist for some of them, but not for most. Treatments exist for some but not all. Many are hard to diagnose in initial stages but deadly in later stages. The World Health Organization struggles with the problem of managing the epidemiology of a large number of diseases over a huge geographic region with many people in it, few of which have resources to pay the cost of a vaccine or a cure (if either exists).

It is probable that disease played a major role in disrupting early attempts to colonize or even trade with African peoples. In 1805 Mungo Park took 44 Europeans on an expedition to discover the source of the Niger River. All but 4 died. In 1816 James Tuckey tried to explore the Congo and lost 37% of his company. In 1832 M'Gregor Laird took 48 men on two ships to explore the Niger delta. Nine survived. In 1841 Trotter took ships up the Niger carrying a mixed crew including 145 Caucasian Europeans and 133 Africans from Sierra Leone: 130 of the Caucasians got sick and 50 died, but none of the African crew got sick. In every one of these examples the explorers reported how disease took the lives of their company. In these instances the diseases of the area actually protected its inhabitants from (largely) unwanted trade or colonization.

It is likely that modern human beings existed first in East Africa. The famous Leakey fossils represent one kind of archeological evidence. Fossil evidence is now backed by DNA studies that demonstrate a higher level of genetic diversity in current African populations than elsewhere, which is generally considered evidence of a longer evolutionary period for the species. It is reasonable to expect that the cradle of humanity would also be the cradle of its most tenacious and adaptable diseases. People who live in colder areas are fortunate in this regard as many of these diseases are highly specialized to make use of insect vectors or animal reservoirs that are only present in warmer areas. Projections of global warming, however, indicate that the region affected by all of these diseases is likely to grow.

13.1. Biological context

In this chapter we will discuss the pharmacokinetics of an anti-malarial drug and in doing so describe the modeling process from start to finish. An excellent reference for general information about drugs used to combat tropical diseases is [1]. Each chapter in this book corresponds to a drug and summarizes what is known (if anything) about the pharmacokinetcs of the drugs. Each chapter also includes an excellent bibliography.

The four species of malaria are *Plasmodium vivax*, *Plasmodium malariae*, *Plasmodium ovale* and *Plasmodium falciparum*. The first drug we consider is halofantrine, an antimalarial drug developed by the US military that is effective against the fourth species mentioned above. The standard antimalarial drug chloroquine is effective against the first three species but the fourth species is resistant to that drug. Halofantrine is administered orally and its pharmacokinetics has been studied extensively [2, 3, 4, 5]. It is the only drug effective against *P. falciparum* but should not be taken to prevent the contraction of the disease.

Halofantrine has been found to be well-tolerated by patients and effective against the multidrug-resistant *P. falciparum* strain of malaria when 500 mg was administered every 6 hours for 3 doses. It is safe for both children and adults and has a cure rate of between 80% and 100%. If, in addition, another course of three doses is given to the non-immune starting on day 7, the cure rate becomes essentially 100%. The side effects in humans of the drug are mild and transient: they include nausea, diarrhea, vomiting abdominal pain, rash, pruritus.

13.2. Multi-compartment pharmacokinetic models

In Chapter 11 we talked about the pharmacokinetics of a drug that can be modeled as being distributed into a single compartment. In Chapter 12, we have already seen models that proposed that the drug goes into parts of the body other than the circulatory system. So, we think it would be helpful to describe the features of such a model in a little more generality.

First, it would be natural to ask how the need for a second compartment can be determined. If the drug really were modeled by a single compartment we could concentration curves as in Figure 13.1. Data that should be modeled with a one-compartment model exhibits the following property: the semilog plot of the data is linear.

Implicit in our description so far is that the signature shape of the concentration curve for data modeled with a two-compartment model should not be linear. What is often seen in real world data is an initial deviation from the straight line on the semilog plot but becomes linear as time goes on. See Figure 13.2 for an example of such a plot. In this section we will discuss how to model drugs that appear to have more than one compartment. In particular, we assume that one compartment is

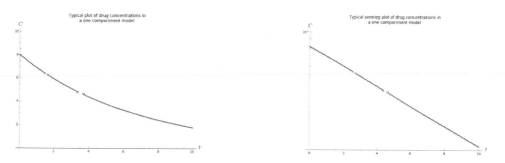

FIGURE 13.1. The image on the left is the plot of the solution to the IVP $C' = -0.15C$, $C(0) = 8$. It is an example of exponential decay. The image on the right is the semilog plot of the same curve. In the latter case we get a straight line, consistent with the fact that we have exponential decay.

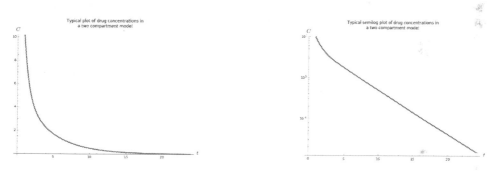

FIGURE 13.2. The image on the left is the plot of the solution to an IVP for a two-compartment model with an IV bolus dose. The image on the right is the semilog plot of the same curve. In this second image we see that there is an initial quick decay and then a long term linear decay.

the central compartment which is the one that includes the circulatory system and is the one that we can most easily measured in the lab. The second compartment is the peripheral compartment and consists of bodily tissues. As such, it is much harder to measure the concentration of the drug in this second compartment and so our main interest is in finding concentrations in the central compartment.

Such a multi-compartment pharmacokinetic model for a drug delivered intravenously can be modeled visually as in Figure 13.3. The system depicted there can be modeled via a system of two linear differential equations. For each compartment, the verbal description of what is going on would be:

rate of change in compartment X = (rate into X) − (rate out of X).

So, we get the following system:

$$\frac{dX_1}{dt} = = (\text{rate into central compartment}) - (\text{rate out of central compartment})$$
$$= +k_{21}X_2 - k_{el}X_1 - k_{12}X_1$$

$$\frac{dX_2}{dt} = (\text{rate into tissue compartment}) - (\text{rate out of tissue compartment})$$
$$= +k_{12}X_1 - k_{21}X_2.$$

We assume the rates into and out of the compartments are proportional to the quantity of drug in each compartment like in Chapters 11 and 12: a drug's likelihood to move across a membrane is proportional to the number of molecules there are of that drug.

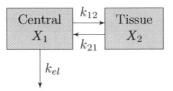

FIGURE 13.3. A two-compartment open model for the pharmaco-kinetics of a drug delivered intravenously.

The system can be solved analytically in a variety of ways (see Chapter 14 for how to solve this system numerically and Chapter 12 for how to do this in Sage). When that is done, we get the following function for the concentration $C_1(t)$ of the drug in the central compartment:

$$C_1(t) = Ae^{-\alpha t} + Be^{-\beta t}$$

where

$$\alpha + \beta = k_{el} + k_{21} + k_{12}$$
$$\alpha\beta = k_{el}k_{21}.$$

EXERCISE 13.1. Show α and β are the roots of the polynomial

$$x^2 - (\alpha + \beta)x + \alpha\beta.$$

Use the quadratic formula to write down expressions for α and β in terms of k_{el}, k_{12} and k_{21}.

Many pharmacokinetic papers report the apparent volume of distribution V_d of the central compartment (or the ratio of V_d to dose or the ratio of V_d to the bioavailability the dose FDose). Using the values of the parameters computed in Exercise 13.1 and the known dose and volume of distribution, we can compute

$$A = \frac{(\text{Dose})(\alpha - k_{21})}{V_1(\alpha - \beta)}$$
$$B = \frac{(\text{Dose})(k_{21} - \beta)}{V_1(\alpha - \beta)},$$

where we assume $\alpha > \beta$.

EXERCISE 13.2. Use the result of Exercise 13.1 to answer this question: A drug follows first order (i.e., linear) two-compartment pharmacokinetics. After looking in the literature we find a number of parameter values for this drug. These numbers represent the microconstants (V_1, k_{el}, k_{12} and k_{21}) for this drug. In order for us to be able to calculate the drug concentration after a single IV bolus dose these parameters need to be converted into values for the macroconstants. The k_{el} and V_1 for this drug in this patient (90.5 kg) are $0.192\mathrm{hr}^{-1}$ and 0.39 L/kg, respectively. The k_{12} and k_{21} values for this drug are 1.86 and 1.68 hr^{-1}, respectively. What is the plasma concentration of this drug 1.5 hours after a 500 mg, IV Bolus dose. In order to complete this calculation first calculate the appropriate A, B, α and β values.

13.3. Method of residuals

All the parameters mentioned in the previous section (and more) can be computed knowing A, B, α and β. The idea is that we start with concentrations collected from a patient experimentally and first determine β and B and the α and β. Recall the concentration of the drug in the central compartment at time t is given by

$$C_1(t) = Ae^{-\alpha t} + Be^{-\beta t},$$

with $\alpha > \beta$.

EXERCISE 13.3. Justify the following claim. Consider the tail seen in the semi-log plot of Figure 13.2. For large t, the tail is a line with slope $-\beta$ and y-intercept B.

The line referenced in Exercise 13.3 can be computed using linear regression. See Chapter 11 for Sage code to compute the slope and intercept of this line via linear regression.

EXERCISE 13.4. Now that we have β and B we consider the data at the beginning of the concentration versus time plot. Justify the claim that these data are given by $C_{\mathrm{early}}(t) = y(t) - Be^{\beta t}$ where $y(t)$ is the observed calculation at time t. Justify the second claim that if plotted on a semilog scale, the graph $C_{\mathrm{early}}(t)$ is linear, with slope $-\alpha$ and intercept A.

By means of Exercises 13.3 and 13.4 we can compute A, B, α and β from observed data. With the values of these parameters, we can conclude the following:

$$V_1 = \frac{\text{Dose}}{A + B}$$
$$k_{21} = \frac{A\beta + B\alpha}{A + B}$$
$$k_{el} = \frac{\alpha\beta}{k_{21}}$$
$$k_{12} = \alpha + \beta - k_{21} - k_{12}.$$

EXERCISE 13.5. Use the equations in this chapter and previous ones to justify the above equations. Be sure to check units.

EXERCISE 13.6. If the values of A, B, α and β are either known or calculated, describe a method that would allow you to figure out what Dose should be administered to a patient as a single IV bolus if you want to achieve a desired concentration at a desired time. Hint: You know the concentration at a second time, too.

13.4. The model for halofantrine

The mechanism of action of halofantrine is not known. In such cases, when a drug is administered orally a first approximation is to assume that after absorption the drug distributes rapidly in the central compartment consisting of blood, extracellular fluid and highly perfused tissues. After distributing in the central compartment, it then distributes in the tissue compartment in which the drug stabilizes less quickly and consists of poorly perfused tissues. The drug is eliminated through the central compartment. A visual representation of the model can be seen in Figure 13.4. We point out its visual similarity to the model for lead in the body introduced in Chapter 12 but point out that we use the standard variables for the rates between compartments and the volumes and concentrations in the compartments.

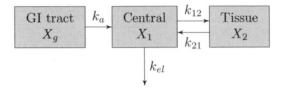

FIGURE 13.4. A two-compartment open model for the pharmacokinetics of halofantrine in the human body.

This leads us to be able to write down a system of differential equations, one equation for each compartment. The differential equation for a give compartment is of the form:

rate of change in compartment $X =$ (rate into X) $-$ (rate out of X)

where we assume the rates between compartments in Figure 13.4 are constant. So in particular, we get this system:

(13.1)
$$\frac{dX_g}{dt} = (\text{ rate into gut }) - (\text{ rate out of gut })$$
(13.2) $= 0 - k_a X_g$

$$\frac{dX_1}{dt} = (\text{ rate into central compartment }) - (\text{ rate out of central compartment })$$
$$= +k_a X_g + k_{21} X_2 - k_{el} X_1 - k_{12} X_1$$

$$\frac{dX_2}{dt} = (\text{ rate into tissue compartment }) - (\text{ rate out of tissue compartment })$$
$$= +k_{12} X_1 - k_{21} X_2.$$

What would the plasma concentrations over time look like for drug that is administered orally and then distributed via a two-compartment model? At the beginning, as the drug is absorbed from the gut into the central compartment the concentration in the blood increases. Once some of the drug is absorbed, a two-phase process begins. The first phase is the distribution phase and exhibits a very

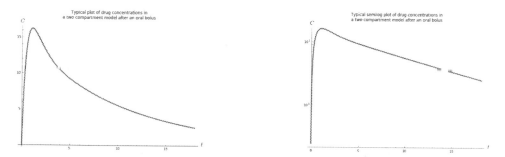

FIGURE 13.5. The image on the left is the plot of the solution to an IVP for a two-compartment model with an oral bolus dose. The image on the right is the semilog plot of the same curve. In this second image we see that there is an initial quick absorption from the gut, a quick decay and then a long term linear decay.

fast decay as the drug is being distributed to the tissue compartment and eliminated. This goes on until a pseudo-equilibrium is reached; i.e., until the influx into and efflux out of the tissue are equal. After this pseudo-equilibrium is reached, the second phase, known as the elimination phase begins. In this phase both elimination and redistribution occur. See Figure 13.5 for a picture of what a typical graph of plasma concentration versus time looks like.

The concentration $C_p(t)$ in the central compartment at time t is given by a function of the form

$$C_p(t) = Ae^{-\alpha t} + Be^{-\beta t} + Ce^{-k_a t},$$

where the α, β, $-k_a$, A, B and C can all be computed by the method of residuals. A more explicit version of this formula is:

$$(13.3) \qquad C_p(t) = \left(\frac{k_a F \text{Dose}}{V_d} \right) \left(\frac{(k_{21}-\alpha)e^{-\alpha t}}{(\beta-\alpha)(k_a-\alpha)} + \frac{(k_{21}-\beta)e^{-\beta t}}{(k_a-\beta)(\alpha-\beta)} + \frac{(k_{21}-k_a)e^{-k_a t}}{(\alpha-k_a)(\beta-k_a)} \right).$$

Here F is the bioavailability of the drug; that is the fraction of the drug that is absorbed by the body and V_d is the volume of distribution as defined earlier.

EXERCISE 13.7. Keeping in mind that $V_d = X_1(t)/C_p(t)$, verify that (13.3) is really a solution to (13.1) (after identifying a reasonable initial condition).

The question is how we find these data in the literature. Papers in pharmacokinetics often report the dose, $t_\alpha^{1/2}$, $t_\beta^{1/2}$, V_d/F, $t_{\max}$, $C_{\max}$ and AUC_{0-t} (for some t) and we are left to compute α, β, k_a and k_{21}. The first two half-lives allow for the computation of α and β, leaving us with two parameters left: k_a and k_{21}. To find the values of these parameters, we need to find a system of two equations in two unknowns. The first one is provided by

$$C_{\max} = C_p(t_{\max}).$$

The other is a little more complicated but can be determined based on the AUC value. AUC stands for Area Under the Curve and is just the sum of the areas of the trapezoids under the data points computed in the lab. See Figure 13.6 for a visual description of what we mean. We want to extrapolate this AUC computed based on data up to some finite time t to infinity because we know the AUC to infinity of $C_p(t)$.

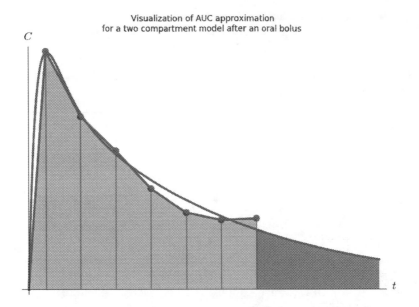

C

Visualization of AUC approximation
for a two compartment model after an oral bolus

t

FIGURE 13.6. This image illustrates how one computes AUC. For an AUC based on plasma samples, one just adds up the areas of the trapezoids determined by the data aand the x-axis. This is the area depicted on the left of the image. The image on the right is the AUC under the tail for times after the last sample was taken. We approximate $AUC_{0-\infty}$ by adding these two areas together.

We describe how to compute the latter first. The following is an example of an improper integral:

$$\int_0^\infty Ae^{-\alpha t}\ dt := \lim_{b\to\infty} \int_0^b Ae^{-\alpha t}\ dt.$$

We compute this as follows:

$$\begin{aligned}
\int_0^\infty Ae^{-\alpha t}\ dt &= \lim_{b\to\infty} \int_0^b Ae^{-\alpha t}\ dt \\
&= \lim_{b\to\infty} A\frac{-1}{\alpha}e^{-\alpha t}\big|_{t=0}^b \\
&= \lim_{b\to\infty} \frac{-A}{\alpha}\left[e^{-\alpha b} - e^0\right] \\
&= \frac{A}{\alpha}
\end{aligned}$$

because $e^{-\alpha b} \to 0$ and $b \to \infty$. Thus the AUC of $C_p(t)$ from 0 to infinity will be

$$AUC_{0-\infty}(C_p) = \frac{A}{\alpha} + \frac{B}{\beta} + \frac{C}{k_a}.$$

Now, if we can compute this AUC in a different way, we will get a second equation. We can break up $AUC_{0-\infty}$ as $AUC_{0-T} + AUC_{T-\infty}$ where the first AUC is computed based on the data and

$$AUC_{T-\infty} \approx \int_T^\infty Be^{-\beta t}\ dt$$

because as $t \to \infty$ the elimination term is the dominant term (the effects of the other two terms are negligible for large t).

EXERCISE 13.8. Show

$$\int_T^\infty Be^{-\beta t} \, dt = \frac{B}{\beta} e^{-\beta T}.$$

Using the result of this exercise, we get a second equation in terms of k_{21} and k_a and we can solve these equations numerically.

EXERCISE 13.9. In [4] the following values are reported in Table 1:
$V_d/F = 100$ L/kg $t_{1/2\alpha} = 1.9$ hr $t_{1/2\beta} = 43.2$ hr
$t_{\max} = 2.8$ hr $C_{\max} = 331$ μg/L $AUC_{0-100} = 82.4$ mg h/L
Find k_{21} and k_a using the method described above.

13.5. Problems

PROBLEM 13.10. In [2] the mean halofantrine data are computed for 12 patients. The values computed are roughly:

(0,0)
(0.5, 20)
(1, 195)
(2, 390)
(3, 700)
(4, 715)
(6, 620)
(6.5, 580)
(7, 470)
(8, 590)
(9, 705)
(10, 870)
(12, 700)
(12.5, 710)
(13, 810)
(14, 790)
(16, 1150)
(18, 1120)
(22, 850)
(24, 610)
(48, 370)
(72, 170)
(96, 150)
(168, 70)
(336, 60)

The time is given in hours and the concentrations are given in ng/mL.

Use these data and the methods developed in this chapter to find t_{max}, C_{max}, $t_\alpha^{1/2}$, $t_\beta^{1/2}$, and AUC. Use these constants to predict at what time the concentration is expected to dip below 25 ng/mL.

PROBLEM 13.11. What role do the size of the constants k_{12} and k_{21} have on the shape of the concentration versus time curve? In particular, explore what happens when

- $k_{12} \approx k_{21} \ll k_a$;
- $k_{12} \approx k_{21} \gg k_a$;
- $k_{12} \gg k_{21} \gg k_a$;
- $k_{12} \gg k_a \gg k_{21}$;
- $k_{21} \gg k_a \gg k_{12}$; and
- $k_{21} \gg k_{12} \gg k_a$.

Additionally, explore what happens in other cases discover something about what role the relative sizes of k_{12}, k_{21}, k_a and k_{el} have on C_{max} and the half-lives. (Here $\gg$ means "much greater than" and "much greater than" might be something like an order of magnitude or two. Similarly for $\ll$.)

Bibliography

[1] Urban Hellgren, Orjan Ericsson, Yakoub AdenAbdi, and Lars L Gustafsson. *Handbook of drugs for tropical parasitic infections.* CRC Press, 1995.

[2] J Karbwang, KA Milton, K Na Bangchang, SA Ward, G Edwards, and D Bunnag. Pharmacokinetics of halofantrine in Thai patients with acute uncomplicated *falciparum* malaria. *British Journal of Clinical Pharmacology*, 31(4):484–487, 1991.

[3] J Karbwang, SA Ward, KA Milton, Na K Bangchang, and G Edwards. Pharmacokinetics of halofantrine in healthy Thai volunteers [letter]. *British Journal of Clinical Pharmacology*, 32(5):639–640, 1991.

[4] Juntra Karbwang and Kesara Na Bangchang. Clinical pharmacokinetics of halofantrine. *Clinical Pharmacokinetics*, 27(2):104–119, 1994.

[5] KA Milton, G Edwards, SA Ward, ML Orme, and AM Breckenridge. Pharmacokinetics of halofantrine in man: effects of food and dose size. *British Journal of Clinical Pharmacology*, 28(1):71–77, 1989.

CHAPTER 14

Euler's Method for Systems of Differential Equations

In the preceding chapters we have seen IVPs of the form

$$\frac{dx}{dt} = f(x, y)$$

$$\frac{dy}{dt} = g(x, y)$$

with initial condition $x(0) = x_0$ and $y(0) = y_0$. Here x has been the amount of drug, say, in the first compartment and y the amount of drug in, say, the second compartment. The function $f(x, y)$ is the difference between the rate of drug into the first compartment and the rate of drug out of the first compartment–for our context, the functions depend on the amounts in the compartments (x and y) and have no explicit dependence on time. As we say in the preceding chapters, when the system is simple enough (in particular, linear and homogeneous) we can find explicit solutions to the IVP. In what follows, in predator-prey modeling in particular, we will be looking at nonlinear systems with explicit time dependence and so it will be useful to have a method to solve IVPs numerically. See [1] for a textbook that has an early description of this method.

In a similar way to Euler's method in Chapter 6, we consider an IVP of the form

$$\frac{dx}{dt} = f(t, x, y)$$

$$\frac{dy}{dt} = g(t, x, y)$$

with initial condition $x(0) = x_0$ and $y(0) = y_0$. Suppose, for example, our IVP is

$$\frac{dx}{dt} = x(1 - y)$$

$$\frac{dy}{dt} = y(x - 1)$$

with $x_0 = 6$ and $y_0 = 3$. We are trying to find the curve that starts here and satisfies the differential equation. So we have to ask where do we end up after a small amount of time Δt passes. The differential equation tells us which way to move and the Δt tells us for how long to walk.

So, in our case, if $\Delta t = 0.1$, we see that

$$x(\text{new time}) \approx (\text{current } x \text{ value}) + (\text{time interval})(\text{rate of change of } x \text{ at current time})$$

$$x(0.1) \approx x(0) + (0.1) \left. \frac{dx}{dt} \right|_{x=6, y=3}$$

$$= 6 + (0.1)(6(1 - 3))$$
$$= 4.8.$$

Similarly, for y, we have

$$y(\text{new time}) \approx (\text{current } y \text{ value}) + (\text{time interval})(\text{rate of change of } y \text{ at current time})$$

$$y(0.1) \approx y(0) + (0.1)\left(\frac{dy}{dt}\bigg|_{x=6,y=3}\right)$$
$$= 3 + (0.1)(3(6 - 1))$$
$$= 4.5.$$

We would proceed in this way to find more approximations of $x(t)$ and $y(t)$ for several values of t. Then, we would plot the resulting values of $x(t)$ and $y(t)$.

EXERCISE 14.1. Continue this process for the example IVP above and make a plot of $x(t)$ and one of $y(t)$ for $0 \le t < 2$.

14.1. Vector notation

In this section we will use vector notation. A vector is nothing more than a way to organize a list of numbers or functions and the main advantage for us is that vectors can be used to consider systems of an arbitrary number of dependent variables. Suppose we have n compartments and let x_i denote the amount of drug in the ith compartment. Then we could derive a system such as

$$(14.1) \qquad \frac{dx_1}{dt} = f_1(t, x_1, \ldots, x_n)$$

$$(14.2) \qquad \frac{dx_2}{dt} = f_2(t, x_1, \ldots, x_n)$$

$$\vdots$$

$$\frac{dx_n}{dt} = f_n(t, x_1, \ldots, x_n)$$

as we have done in the previous chapters. Notationally this gets awkward and clumsy pretty quickly.

Instead, we define

$$\mathbf{X}(t) = \begin{pmatrix} x_1(t) \\ x_2(t) \\ \vdots \\ x_n(t) \end{pmatrix}$$

and

$$\frac{d\mathbf{X}}{dt} = \mathbf{X}'(t) = \begin{pmatrix} x_1'(t) \\ x_2'(t) \\ \vdots \\ x_n'(t) \end{pmatrix}.$$

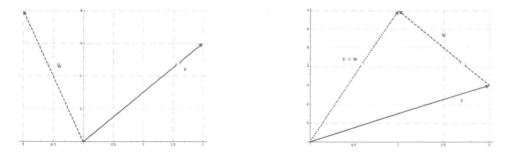

FIGURE 14.1. The figure on the left shows the two vectors $\mathbf{v} = \left(\begin{smallmatrix} 2 \\ 3 \end{smallmatrix}\right)$ and $\mathbf{w} = \left(\begin{smallmatrix} -1 \\ 4 \end{smallmatrix}\right)$. The figure on the right shows how to visualize the sum of vectors as a composition of displacements: by moving the tail of vector $\mathbf{w}$ to the tip of vector $\mathbf{v}$ and drawing the vector $\mathbf{v} + \mathbf{w}$ from the tail of $\mathbf{v}$ to the tip of vector $\mathbf{w}$, yields the vector that represents the overall motion given by first moving along $\mathbf{v}$ and then along $\mathbf{w}$.

With this notation the left-hand side of (14.1) becomes $\frac{d\mathbf{X}}{dt}$. Similarly, we let

$$\mathbf{F}(t, x_1(t), \ldots, x_n(t)) = \mathbf{F}(t, \mathbf{X}(t)) = \begin{pmatrix} f_1(t, x_1(t), x_2(t), \ldots, x_n(t)) \\ f_2(t, x_1(t), x_2(t), \ldots, x_n(t)) \\ \vdots \\ f_n(t, x_1(t), x_2(t), \ldots, x_n(t)) \end{pmatrix}.$$

Then with this notation (14.1) becomes

$$\frac{d\mathbf{X}}{dt} = \mathbf{F}(t, \mathbf{X}(t))$$

which is clearly much easier to write. We indicate an initial condition by

$$\mathbf{X}_0 = \begin{pmatrix} (x_1(t))_0 \\ (x_2(t))_0 \\ \vdots \\ (x_n(t))_0 \end{pmatrix}.$$

Now, what does Euler's method mean in this case? Vectors are not just lists of numbers, it also makes sense to add two vectors. If we take one vector $\mathbf{x} = \begin{pmatrix} x_1 \\ \vdots \\ x_n \end{pmatrix}$ and another $\mathbf{y} = \begin{pmatrix} y_1 \\ \vdots \\ y_n \end{pmatrix}$ their sum is $\mathbf{x} + \mathbf{y} = \begin{pmatrix} x_1 + y_1 \\ \vdots \\ x_n + y_n \end{pmatrix}$. That is, we add them component by component. That is the algebraic meaning of the sum, but what does it mean geometrically? If we think of a vector as a displacement from the origin, then the sum of two vectors $\mathbf{v} + \mathbf{w}$ is a displacement by $\mathbf{v}$ followed by a displacement by $\mathbf{w}$. See Figure 14.1 for a visual representation of what the sum means geometrically.

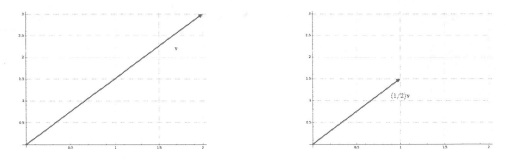

FIGURE 14.2. The figure on the left shows the vector $\mathbf{v} = \left(\begin{smallmatrix} 2 \\ 3 \end{smallmatrix}\right)$. The figure on the right shows how to visualize the multiplication of vector by a number as a scaling: the vector $\frac{1}{2}\mathbf{v}$ has the same direction as $\mathbf{v}$ but is half as long.

In addition to adding two vectors, we can also multiply them by numbers (scalars). If we take a vector $\mathbf{x} = \left(\begin{smallmatrix} x_1 \\ \vdots \\ x_n \end{smallmatrix}\right)$ and multiply it by a (real) number c, we get $c\mathbf{x} = \left(\begin{smallmatrix} cx_1 \\ \vdots \\ cx_n \end{smallmatrix}\right)$. See Figure 14.2 for a geometric interpretation of what this kind of multiplication means.

To translate Euler's method into this notation, we again illustrate what needs to be done via the same example as above. Recall that we had the IVP

$$\frac{dx}{dt} = x(1 - y)$$
$$\frac{dy}{dt} = y(x - 1)$$

with $x_0 = 6$ and $y_0 = 3$. The notation becomes

$$\mathbf{X}(t) = \begin{pmatrix} x(t) \\ y(t) \end{pmatrix}$$
$$\mathbf{X}'(t) = \begin{pmatrix} x(t)(1 - y(t)) \\ y(t)(x(t) - 1) \end{pmatrix}$$
$$\mathbf{X}_0 = \mathbf{X}(0) = \begin{pmatrix} 6 \\ 3 \end{pmatrix}.$$

If once again we consider a time step $\Delta t = 0.1$, then, to compute $\mathbf{X}(0.1)$ we merely compute the following:

$$\mathbf{X}(0.1) \approx (\,\text{current position}\,) + (\,\text{time interval}\,)(\,\text{current rate of change}\,)$$
$$= \mathbf{X}(0) + (\Delta t)(\mathbf{X}'(0))$$
$$= \left(\begin{smallmatrix} 6 \\ 3 \end{smallmatrix}\right) + 0.1 \left(\begin{smallmatrix} 6(1-3) \\ 3(6-1) \end{smallmatrix}\right)$$
$$= \left(\begin{smallmatrix} 6 \\ 3 \end{smallmatrix}\right) + \left(\begin{smallmatrix} (0.1)(-12) \\ (0.1)(15) \end{smallmatrix}\right) = \left(\begin{smallmatrix} 6-1.2 \\ 3+1.5 \end{smallmatrix}\right) = \left(\begin{smallmatrix} 4.8 \\ 4.5 \end{smallmatrix}\right).$$

We would proceed in this way to find $\mathbf{X}(t)$ for other values of t. We point out that the formula just stated is very similar to the formula in Chapter 6. This is another advantage of introducing the vector notation.

EXERCISE 14.2. Continue this process with the example IVP from above and make a plot of $\mathbf{X}(t)$ for $0 \leq t < 2$. Compare your answers here with the answers from Exercise 14.1.

14.2. Existence and uniqueness

Just like in Chapter 6, we have a theorem that essentially guarantees that, as long as the components of the vector $\mathbf{X}'(t)$ are continuous, there is guaranteed to be a solution to the IVP we are considering. In particular,

THEOREM 14.3. *Let*
$$\mathbf{X}'(t) = \mathbf{F}(t, \mathbf{X}(t))$$
be a system of differential equations. Suppose that at time t_0 we have an initial condition of $\mathbf{X}_0$. Suppose that the function $\mathbf{F}$ is continuously differentiable. Then there is an $\varepsilon > 0$ and a function $\mathbf{X}(t)$ defined for $t_0 - \varepsilon < t < t_0 + \varepsilon$ such that $\mathbf{X}(t)$ satisfies the IVP $\mathbf{X}'(t) = \mathbf{F}(t, \mathbf{X}(t))$ and $\mathbf{X}(t_0) = \mathbf{X}_0$. Moreover, for t in this interval, this solution is unique.

So, as long as our stepsize keeps us within the region where $\mathbf{X}(t)$ is continuous and defined, we will end up with a good approximation to the solution. For us in this book, the functions $\mathbf{F}$ will be continuous and differentiable everywhere. Thus a reasonably small stepsize will give a reasonably good approximation to our solution.

14.3. Problems

PROBLEM 14.4. In Chapter 6, we explained the formula for Euler's method for a single differential equation in a particular way involving antiderivatives. Re-explain Euler's method for a single differential equation in a way analogous to our description in this chapter.

PROBLEM 14.5. Consider the system
$$\frac{dx}{dt} = 2x$$
$$\frac{dy}{dt} = y$$
with initial condition $x(0) = 1$ and $y(0) = 3$.

(1) Write the system in vector notation, using $\mathbf{X}(t) = \begin{pmatrix} x(t) \\ y(t) \end{pmatrix}$.
(2) Check that the curve $\mathbf{X}(t) = \begin{pmatrix} e^{2t} \\ 3e^t \end{pmatrix}$ is a solution to the initial value problem.
(3) Use Euler's method with step size $\Delta t = 0.5$ to approximate this solution, and check how close the approximate solution is to the real solution when $t = 2$, $t = 4$, and $t = 6$.
(4) Use Euler's method with step size $\Delta t = 0.1$ to approximate this solution, and check how close the approximate solution is to the real solution when $t = 2$, $t = 4$, and $t = 6$.
(5) How and why do the Euler approximations differ from the solution?

Bibliography

[1] William E Boyce. *Differential Equations: An Introduction to Modern Methods and Applications.* John Wiley & Sons, 2010.

CHAPTER 15

Modeling Interlude: Sensitivity Analysis

The models we have considered up to now have treated their parameters as being fixed. Sometimes we have pointed out that some of the parameters fall within some range (e.g., $k_{el} = 0.77 \pm 0.12$). This in turn leads to some range of the values of the dependent variables. In this chapter we ask the more subtle question: suppose there is some output from the model we care about most. Maybe, for example, in the pharmacokinetic models from the preceding chapters, we want to know about C_{max}. In particular, we want to understand which of the following parameters have the biggest effect on C_{max}: does k_{el} affect it the most? the size of the dose? the value of k_a?

Sensitivity analysis is the systematic investigation of the model's response to either

(1) the perturbation of the model's quantitative components (e.g., initial conditions, parameter values, etc.), or

(2) the variation of the model's qualitative factors (e.g., number of compartments, the existence or not of arrows between certain compartments, etc.).

In this chapter, we focus more on the first. We present a method of sensitivity analysis that is empirical in the sense that it requires evaluating the model many times.

15.1. Notation

In the previous chapter, in which we described Euler's method for systems of differential equations, we introduced the vector notation for the system. Namely,

$$x'(t) = f(t, x, y)$$
$$y'(t) = g(t, x, y) \qquad \Rightarrow \mathbf{X}'(t) = \mathbf{F}(t, \mathbf{X}), \mathbf{X}'(0) = \mathbf{X}_0$$
$$x(0) = x_0, y(0) = y_0.$$

Now we are going to treat the parameters (the k_a, k_{el}, Dose, etc.) as variables. We will put them in a vector

$$\mathbf{p} = \begin{pmatrix} p_1 \\ p_2 \\ \vdots \\ p_k \end{pmatrix},$$

where each the variables $p_1, \ldots, p_k$ correspond to the parameters in the system of differential equations (e.g., maybe p_1 is k_a, p_2 is k_{el}, etc.). To make the IVP's dependence on the parameters explicit, we will write the system in vector form as

$$\mathbf{X}'(t) = \mathbf{F}(t, \mathbf{X}, \mathbf{p}).$$

% change in	$a+5\%$	$a-5\%$	$e+5\%$	$e-5\%$	$d+5\%$	$d-5\%$	$b+5\%$	$b-5\%$
$C_{\max}$	$+5$	-5	0	0	0	0	-4.8	$+5.3$

TABLE 15.1. In each column a parameter in the lead model describes above is either increased or decreased by 5% and the value report is the percent increase or decrease of $C_{\max}$.

Unless stated otherwise, we assume $\mathbf{X}(t)$ has n components denoted by $x_1(t)$, $x_2(t),\ldots, x_n(t)$ and k parameters denoted by $p_1,\ldots,p_k$.

In general, sensitivity analysis is about the percent change in dependent variable x_i compared to the percent change in parameter p_j. That is, the sensitivity s_{ij} of the ith dependent variable with respect to the jth parameter is

$$s_{ij} = \frac{\text{percent change in } x_i}{\text{percent change in } p_j}$$

$$= \frac{(\text{change in } x_i)/x_i}{(\text{change in } p_j)/p_j}$$

$$= \frac{\Delta x_i / x_i}{\Delta p_j / p_j}$$

$$= \frac{\Delta x_i}{\Delta p_j} \cdot \frac{p_j}{x_i}$$

where the last step follows from how you divide fractions. We present this notation to show how general the framework for sensitivity analysis can be and to point out that using linear algebra one can calculate sensitivities with a few simple calculations. The method we present, though, is much more numerical.

15.2. The basic question and a basic solution

Suppose we are considering the pharmacokinetic model for lead described earlier. As can be seen in Figure 12.1 we have parameters a, e, d and b and we want to know how much $C_{\max}$ changes as we change each of the parameters. In particular, we might ask by what percent does $C_{\max}$ change if we change the parameter a by 5%. Well, the simplest approach is to do just that. We run the model with particular values of the parameters a, e, d and b, and find $C_{\max}$ for those values. Then, we increase a by 5% and see how $C_{\max}$ changes, decrease a by 5% and see how $C_{\max}$ changes, increase b by 5% and see how $C_{\max}$ changes, etc.

See Figure 15.1 for a visual representation of a comparison between the model's sensitivity after varying each of the parameters a, e, d and b by 5%. Such a diagram is called a *tornado diagram*. To fix ideas as much as possible, we invent the following values for the parameters in the lead model. Let

$$a = 1000 \qquad e = 7 \qquad d = 2 \quad b = 3$$
$$x_1(0) = 13 \quad x_2(0) = 0.5$$

In this case $C_{\max}$ is 333.3 mg/L. Say we increase the intake of air from the atmosphere by 5% to 1050. Now $C_{\max}$ is 350, an increase of 5%. Similarly, if we decrease it by 5% to 950, $C_{\max}$ decreases by 5%. What happens if we consider b? If we increase it by 5% to 3.15, we get a $C_{\max}$ of 317.5, a decrease of 4.8%. If we decrease it by 5% to 2.85, we get a $C_{\max}$ of 350.9, an increase of 5.3%. Continuing this way we get Table 15.1.

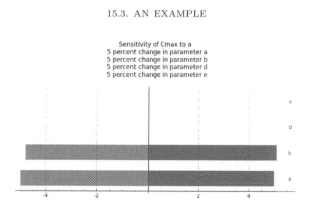

FIGURE 15.1. A tornado diagram of senstivity of the $C_{\max}$ as calculated by the model described in this section. The image shows how the value of $C_{\max}$ responds to a 5% change in each of the four parameters in the model. We point out that one shading of the bars corresponds to an increase in $C_{\max}$ and the other to a decrease.

15.3. An example

We continue with the lead example with the made up values for a, b, d, e and $C(0)$ and $U(0)$. Here's some Sage code:

```
X,Z,t=var('X Z  t')

a=1000

ss = desolve_system_rk4([-(2+3.15)*X+7*Z+a,2*X-7*Z],
    [X,Z],ics=[0,13,.5],ivar=t,end_points=10,step=.1)

XX=[ [i,j] for i,j,l in ss]

LP=list_plot(XX,plotjoined=True,thickness=3,
    legend_label="Mass of lead in bloodstream")

ZZ=[[i,l] for i,j,l in ss]

LP += list_plot(ZZ,color="green",plotjoined=True,
  thickness=3,linestyle="--",
    legend_label="Mass of lead in bones",
      title='Time series of the amount of lead
            in\n two compartments and with constant
            lead inhalation',
        axes_labels=['$t$','$X,Z$'])

show(LP)
```

Notice that we've left an a in the differential equation and that before Sage computes ss, we've set a to be 1000.

If we look at the graph, the equilibrium (which is also the maximal concentration in the central compartment) looks to be about 333.33. To be sure, we write

```
ss[-1]
```

which returns

```
[10.0, 333.333333167084, 95.23809516986218]
```

The returned triple says that at time 10, the concentration in the central compartment was about 333.33 and in the bone was about 95.24.

To see what happens to the equilibrium when we increase a by 5% to 1050.

```
X,Z,t=var('X Z  t')

a=1050

ss = desolve_system_rk4([-(2+3)*X+7*Z+a,2*X-7*Z],
    [X,Z],ics=[0,13,.5],ivar=t,end_points=10,step=.1)

XX=[ [i,j] for i,j,l in ss]

LP=list_plot(XX,plotjoined=True,thickness=3,
    legend_label="Mass of lead in bloodstream")

ZZ=[[i,l] for i,j,l in ss]

LP += list_plot(ZZ,color="green",plotjoined=True,
    thickness=3,linestyle="--",
      legend_label="Mass of lead in bones",
        title='Time series of the amount of lead
              in\n two compartments and with constant
              lead inhalation',
          axes_labels=['$t$','$X,Z$'])

ss[-1]
```

which returns

```
[10.0, 349.9999998251884, 99.99999992825273].
```

From this we see that the equilibrium increased to 350, or increased by 5% (since $350/333.3333\cdots = 1.05$).

We proceed in this way for each parameter, increasing and decreasing each parameter by 5%. In particular, we would next try with $a = 950$. Proceeding in this way we get Table 15.1 and we plot the data in the table using the following Sage function.

```
def tornado_plot(test_var,var_list, pct_list, pos_sensitivity_list,
    neg_sensitivity_list):
    '''

        test_var:
          the variable whose senstivity you are analyzing, a string
        var_list:
          a list of strings that correspond to the parameters you
```

```
                are varying
            pct_list:
              a list of percentages where the ith entry determines how
              much you vary the ith parameter
            pos_sensitivity_list:
              a list of percentages that result when you increase a
              parameter by the appropriate percentage
            neg_sensitivity_list:
              a list of percentages that result when you decrease a
              parameter by the appropriate percentage
      '''

    num_bars = len(var_list)
    P = line([(0,0),(0,num_bars)])
    pos_scale = max(pos_sensitivity_list)
    neg_scale = max(neg_sensitivity_list)
    ss = [" " + str(pct_list[i]) + " percent change in parameter "
            + var_list[i] +"\n" for i in range(num_bars)]
    tt = ''.join(ss)
    P += plot(0,(x,-neg_scale,pos_scale),
            ticks=[ceil(pos_scale/num_bars),[]],
              title="Sensitivity of "+test_var+" to a \n"+tt)
    for i in range(num_bars):
        P += polygon([(0,i+.2),(pos_sensitivity_list[i],i+.2),
                (pos_sensitivity_list[i],i+.8),(0,i+.8)],
                  color=(0,0,1))
        P += polygon([(0,i+.2),(-neg_sensitivity_list[i],i+.2),
                (-neg_sensitivity_list[i],i+.8),(0,i+.8)],
                  color=(1,0,0))
        P += text(var_list[i],(pos_scale*1.1,i+.5))
    return P

PP = tornado_plot('Cmax',['a','b','d','e'], [5,5,5,5],
        [5,5.1,0,0],[5,4.8,0,0])

show(PP,gridlines=True)
```

The plot in Figure 15.1 is generated by the above function and function call.

15.4. Problems

PROBLEM 15.1. Of the various pharmacokinetic constants we would like to calculate from the model for lead represented in Figure 12.1, which do you think will be most sensitive to a small change in the parameters. That is, if we change a by a little, which of the following will change more: $C_{\max}$, $t_\alpha^{1/2}$, $t_\beta^{1/2}$, or AUC? Same question for the other parameters: e_1, e_2, b, d_1, and d_2. Was your instinct correct? Explain.

PROBLEM 15.2. Understand the tornado plot Sage code written above. Try to improve it.

CHAPTER 16

Research Interlude: Writing a Research Paper

The three Ps of scientific research are problem posing, problem solving and peer persuasion. One way scientists use to persuade their peers is by writing papers. Once they have posed and solved a problem, they try to disseminate their results, sometimes in the form of a written paper.

16.1. Components of a scientific paper

In science papers, there is a standard list of sections with a standard order: (1) Title, (2) Authors, (3) Abstract, (4) Introduction, (5) Description of model, (6) Methods, (7) Results and Conclusions, (8) Discussion, (9) Acknowledgments, (10) Bibliography, and (11) Appendix. In what follows we touch on each of the 11 parts.

Title. This is your paper's first impression on a potential reader. It needs to strike a balance between being explicit enough and being too jargon-filled. The most effective titles are ones that summarize the results: A three-dose schedule of nifurtimox is recommended in patients with compromised kidneys.

Authors. Who gets listed and in what order is tricky business. In mathematics, the authors are always listed in alphabetical order of last name. We endorse this convention. In other disciplines, the person who did the bulk of the work is listed as the first author, the person whose lab or research group the work was done in is listed as the last author, and other people who made substantial contributions to the final product are listed in between in some (seemingly arbitrary) order.

Who should be included in a list of authors is also tricky, but there are some pretty clear guidelines. According to the International Committee of Medical Journal Editors [1], a person should be listed as an author if and only if they qualify as follows. A person gets an authorship credit only if conditions (1), (2), (3) and (4) are met:

(1) Substantial contributions to the conception or design of the work; or the acquisition, analysis, or interpretation of data for the work; AND
(2) Drafting the work or revising it critically for important intellectual content; AND
(3) Final approval of the version to be published; AND
(4) Agreement to be accountable for all aspects of the work in ensuring that questions related to the accuracy or integrity of any part of the work are appropriately investigated and resolved.

Abstract. An abstract is a 100-250 word preview of your article and should strike a balance between getting the reader excited to read your article and convincing the reader that the work is serious. It should be more technical than the title and less than the article itself. Abstracts are often stored in databases that are searchable and so you want to be sure you include the keywords of your project.

Being that an abstract is so short, there is a certain art to writing them. We typically write a summary in our words and then iteratively prune it down by removing non-essential words until it is within the word limits.

In general, it is not advisable to use abbreviations (unless they are so standard they are a keyword in their own right) and it is essentially forbidden to use citations.

Introduction. This section should be written with a lay-scientist in mind. Not someone who is an expert in the narrow part of the field that your paper is about, but someone who maybe works in areas related to the focus of the paper. This section needs to include relevant background information and the statement of the problem you are solving. The problem should then be given context by connecting it to the literature and then your solution should be presented and a brief comparison of your solution to existing ones should also be carried out.

Description of model. As discussed in Chapter 7 a verbal, algebraic and graphical representation of the model should be provided. All the variables should be clearly described and a table of the values of parameters (and references where the values can be found) should be provided. The idea is to describe the model so carefully that the reader can reproduce your results and validate the assumptions you have made in your model.

Methods. For models as described in this text, this part will be relatively short. It might say include what numerical method for solving the differential equations was used as well as what tools were used to analyze the output of the numerical method.

Results and Conclusions. It is in this section where the model presented earlier is analyzed physically and graphically. The section includes graphs of solutions of the model, phase portraits of the model. This section should not include an interpretation of the model, its solutions or its qualitative behavior–that is all done in the subsequent section. Here you are just presenting the data and giving the reader some space to do their own preliminary interpretation. A quick comment on graphs and tables: make sure they are carefully labelled (axes, a title, a key, etc.) and also make sure they are absolutely necessary (if the data they convey can be summarized in a sentence or two, do that instead). In Chapter 25, we will have a longer discussion of graphs and tables.

Discussion. In this section you interpret the data presented in the previous section. Here you describe the value added to the field by your model and by your analysis of the model. Your job is to convince the reader that your interpretations are correct and somehow an improvement (more general? slightly different context? slightly better fit to data?) over previous work. If your results are consistent with previous results, highlight that fact; if your results are unexpected, explain why that might be the case.

End the section with a one-sentence summary of your conclusion, clearly explaining why the reader should care.

Acknowledgments. Not every paper you will write will have this section. It is a place to thank the referee who provided useful comments, the colleague who helped you have a breakthrough, the agency or program that funded the work of an author on the paper, the colleague who read an earlier version of the paper, etc.

This section is optional. You can thank those who either helped with the experiments, or made other important contributions, such as discussing the protocol, commenting on the manuscript, or buying you pizza.

Bibliography. We prefer the following kind of bibliography (it is the kind you have seen in every chapter of this text, including this one!). In the text there are references like [1]. The number in the bracket is tied to a particular reference in the bibliography. The bibliography is listed in alphabetical order. Of course this is only one option, but we find it simple and not distracting.

Appendix. This section is optional. Often the details of a technical computation that would be distracting from the main narrative of the paper if they were included in the middle of the paper are relegated to an appendix. Another common occurrence is that if it is appropriate to have many graphs and tables (e.g., maybe you are investigating the effect of tweaking a parameter on a model) in your results section, you might include a summary of the tables in the Results and Conclusions section and put the individual tables in the appendix. Of course this is up to your discretion.

16.2. Putting pen to paper (or fingers to keyboard)

Now that we have discussed the format a paper might take, we give you some practical advice about the process of actually writing the paper.

Edit your paper. Write some, then re-write, then re-write some more. For research papers, it is not common to go through several completely different drafts before settling on a final version.

Scientific writing. The most important thing in scientific writing is to be both accurate and digestible. The content of your paper is hard enough already. A reader should not have to fight against your writing to understand what you are trying to say.

- Scientific writing often use the same word multiple times in the same sentence. Since scientific terms are so carefully defined it only makes sense to use the same word: all that work was spent on defining the word carefully, so why would you use a different word?
- Write in the active voice. Instead of saying "The model was studied...", say "We studied the model and...".
- Write at a level appropriate to the audience and assume that the audience is neither your co-author nor an expert in your field nor a world-leader in a closely-related field. Assume the reader is a solid scientist who might care about your work.
- Write in the first person. In pure mathematics, we write in the first person plural. We use "we". This suggests that the author is a tour-guide, sharing the wonders of their paper with you.
- Use short words. E.g., instead of "possess" use "has". E.g., instead of "take into consideration" use "consider".

- Avoid "to be". E.g., instead of using "the equilibria of the system were found to be at...", use "the equilibria are...".
- Use a spellchecker but also proofread the paper carefully a few times as spellcheckers have been known to miss a hole (*sic*) lot of things.

Bibliography

[1] International Committee of Medical Journal Editors et al. Recommendations for the conduct, reporting, editing, and publication of scholarly work in medical journals. `http://www.icmje.org`, 2013.

CHAPTER 17

Projects for Pharmacokinetic Modeling

Note to the instructor. The two projects in this section are emblematic of the kind of work that a researcher in biology and/or medicine can carry out based on the material in the course so far. Basically, these projects are about the pharmacokinetics of drugs administered to patients and researchers' attempts to model them. In principle, one could take any publically available paper on the pharmacokinetics of a drug and adjust it. E.g., change the dosing schedule, change the clearance rates (e.g., by assuming the patient has liver damage), change the number of compartments, change the method of administration (oral versus IV), etc.

We recommend that, if this project is the first project being carried out by your students, your students work in pairs or small groups. The peer instruction that occurs in such an assignment pays dividends in future projects to be done individually.

17.1. Projects

We have tried to make the two projects quite explicit but there is some open-endedness that may cause some students some worry.

(1) In this project you will investigate the merits of various dosing strategies for Nifurtimox, a drug that is used in the treatment of both sleeping sickness and Chagas disease. A systematic review of the treatment and diagnosis of Chagas can be found in [1]. The CDC recommends three or four doses per day [2]. You will conduct a basic drug study. A useful paper to consider is [4].

Use Sage to compare what happens over five days with the three-dosage regimen versus the four-dosage regimen. Keep a data set with the following values for each run:

F	V	k_a	k_{el}	Run#
Day	Time	Length of run	starting X	starting C
ending X	ending C	max C	min C	

You will turn in your data tables, along with plots of a few runs. After the first dose, what are the maximum and minimum blood concentrations over this period? Because the constants of absorption and elimination fall in a range, you will need to test the extreme values for these.

The value added by your project will be your answers to these three **research questions**:

(a) Research question: Read up a bit on this drug. Based on your reading and your computational experiments, which dosage regimen do you recommend?

(b) Research question: What makes a bigger difference in your C_{max} and C_{min}? The variation in k_a and k_{el} or the variation in dosage strategy? Design a numerical experiment that tests this. Interpret the results.

(c) Research question: How would impaired kidneys affect the dosage regime you have designed? Use your model to suggest an alternative for such individuals.

After you answer the individual questions, spend some time figuring out how to weave them together into a single coherent paper.

Now **write the paper**. It should have the following sections:

- **Introduction:** including background information and statement of the problem you are solving, along with the solution (your recommendation). The thing to highlight here is what your model is adding to our knowledge about the drug.
- **Description of model:** including box diagram, explanation of equations and the actual equations. Also in this section state your parameter values (F, V, k_a and k_{el}) along with how you got them and the source of the information (properly cited of course).
- **Methods:** a short description of what computer runs you did and how you did them.
- **Results:** a synopsis of your data and what you found.
- **Conclusions:** Here you argue for your specific recommendations.
- **Bibliography:** sources cited, including Sage. Always cite your software.
- **Appendix:** your data set.

Be sure to read your paper for grammatical errors and for good style. Your paper should lead the reader to the conclusion your are trying to make; you need to persuade the reader that your conclusions are correct and interesting.

(2) You will develop a two-compartment model for IV bolus administration of Bortezomid. Use Sage to identify parameters that give a model that matches the data in Papandreou *et al.* [3].

The value added by your paper will be your answers to these research questions:

(a) Research question: What box model and equations best represent the dynamics of this drug in the human body? What constants must be identified from data?

(b) Research question: Based on the description given in the paper, how would you set initial conditions for the doses described in Figure 1 of the paper? Notice the log scale.

(c) Research question: Using one of the dosages in Figure 1 of the paper, find constants that give Sage runs that match the data. NOTE the log scale.

(d) Research question: Do the same for the other dosages. Estimate your constants for each of these runs, as well as averages.

(e) Research question: Now, using the dosages that were actually administered to patients, select low, medium and high dosage rates. What does your model predict for these?

(f) Research question: For each of these three doses, how long does the plasma concentration stay above 50% of maximum? Above 25% of maximum? Above 10% of maximum? And what are these levels?

(g) Research question: The paper describes giving a second dose on day 8. How much is left from the first dose, in blood and soft tissue, at that time, according to your model?

(h) Simulate the second dose. Does anything different happen?

(i) Think about the dosage regime in comparison to potential side effects. Read the paper. Think about whether 8 days is a realistic interval for this therapy.

After you answer the individual questions, spend some time figuring out how to weave them together into a single coherent paper.

Now **write your paper**. It should have the following sections:

- **Introduction:** including background information and statement of the problem you are solving, along with the solution (your recommendation).
- **Description of model:** including box diagram, explanation of equations and the actual equations. Also in this section state your parameter values (F, V, k_a and k_{el}) along with how you got them and the source of the information (properly cited of course).
- **Methods:** a short description of what computer runs you did and how you did them.
- **Results:** a synopsis of your data and what you found. Include illustrative figures.
- **Conclusions:** here you argue for your specific recommendations.
- **Appendix:** your data sets or computer runs.
- **Bibliography:** sources cited, including Sage. Always cite your software.

Be sure to read your paper for grammatical errors and for good style. Your paper should lead the reader to the conclusion your are trying to make; you need to persuade the reader that your conclusions are correct and interesting.

Bibliography

[1] Caryn Bern et al. Evaluation and treatment of Chagas disease in the United States: a systematic review. *JAMA*, 298(18):2171–2181, 2007.

[2] Centers for Disease Control. CDC-Chagas Disease-Resources for Health Professionals-Antiparasitic treatment. https://www.cdc.gov/parasites/chagas/health_professionals/tx.html. Accessed: 12/26/2016.

[3] Christos N Papandreou et al. Phase I trial of the proteasome inhibitor Bortezomib in patients with advanced solid tumors with observations in androgen-independent prostate cancer. *Journal of Clinical Oncology*, 22(11):2108–2121, 2004.

[4] C Paulos, J Paredes, I Vasquez, S Thambo, A Arancibia, and G Gonzalez-Martin. Pharmacokinetics of a nitrofuran compound, Nifurtimox, in healthy volunteers. *International Journal of Clinical Pharmacology, Therapy, and Toxicology*, 27(9):454–457, 1989.

Part 4

Predator prey modeling

CHAPTER 18

Undamped Lotka-Volterra Equations

In 1858, British explorer John Speke described the fishing he saw on Lake Victoria:

> There are very few canoes about here, and those are of miserable construction, and only fitted for the purpose they turn them to–catching fish close to the shore. The paddle the fishermen use is a sort of mongrel between a spade and a shovel. The fact of there being no boats of any size here, must be attributed to the want of material for constructing them. On the route from Kazé there are no trees of any girth, save the calabash, the wood of which is too soft for boat-building.

Over a century later this situation still had not changed much. In 1979, only two percent of the canoes on Lake Victoria had outboard motors. Many of the fishermen in the region do not have the capability to fish in the deeper parts of the lake and without sophisticated equipment they cannot keep fish fresh if they stray far from the shore. For an overview of the relationship between fishing, the Lake and the economic health of the communities around the Lake, see [7].

As of this writing, more than 30 million people in the three countries bordering Lake Victoria (Uganda, Kenya, and Tanzania) are in some way dependent on the lake for their livelihood. These people not only include fishermen but many people directly and indirectly associated with the fishing industry. Without the fishing industry, fish sellers and transporters of fish, canoe builders, net-makers, and people who repair canoes and nets would not have work. Further, owners of small shops and hotels, in addition to construction workers and railway workers, benefit greatly from the large population centered around the lake. Many additional people prosper from the reinvestment of money earned from the lake back in local businesses.

The fishing industry began to experience changes that occurred as early as the 1900s when British brought improved fishing equipment to the local fishermen. Traditionally fishermen had used traps, weirs, baskets, and spears. Flax gill nets were introduced in 1905, non-selective beach seines in the early 1920s, and synthetic fiber gill nets in 1952.

In 1908 a railway was completed from Mombasa (on the coast of East Africa) to Lake Victoria. While it took explorer John Speke nine months and an average missionary six months to reach Lake Victoria without the benefit of a railroad, after 1908 the trip only took 2.5 days. The railway facilitated the creation of a previously impractical export industry. This, in addition to a tremendous increase in population in the new urban centers of the region, led to an increased demand for fish which, coupled with the use of better fishing equipment, led to overfishing, decreasing the amount of fish in the lake. As catch size decreased, fishermen were left with no choice other than to decrease mesh size on their nets which further decreased the fish populations. In the early 1950's fishing effort nearly doubled while yield only increased by about 10 percent. The smaller mesh size meant that fishermen caught smaller fish including a significant number of juveniles. Catching young fish made it even more difficult for the populations to replenish themselves. In the early 1970s the commercial catch was as much as 35 percent immature, and some concluded that it was likely that some of the larger species of Haplochromis had almost been eliminated byoverfishing.

18.1. Biological context

Cichlid fish had once served as the main catch for local fishermen along with catfish, carp, and lungfish. The two endemic tilapia species *Oreochromis esculenta* and *O. variabilis* which had previously been important to the fishing industry were almost completely wiped out by the early 1950s. Although the Lake Victoria fishery should be a renewable resource, it has been exploited to the point that some feel it may not ever be able to return to its original level of productivity.

The cichlid population was in serious decline from overfishing during the same period of time as the algae bloom we investigated in Chapter 3, resulting in a major change for the lake. When the base of the food chain experiences rapid growth (as the algae did) one would expect a corresponding growth in the population of fish feeding on the algae, which include the cichlids. The extra biomass generated by the algae would then be transferred to a biomass of certain types of cichlids, which would then increase the biomass of their predators, and so on up the food chain. When the intermediate link of algae-eaters is removed or drastically reduced (as it was in this case by overfishing), the algae continue to bloom unabated. This has two consequences. First, the density of algae makes it difficult for light to reach the depth of the lake, thereby concentrating the algal and other plant biomass near the surface. Because plants produce oxygen, their concentration near the surface results in less oxygen in the depth of the lake. In fact, lake waters tend to separate into layers by temperature, making it difficult to mix oxygen produced at the surface with the deep waters. So, by their sheer numbers, the algae can change the chemistry of a large body of water. The second effect is the result of the death of a portion of the algae and other plant populations. Because nobody is eating these plants, a certain proportion of them die and drop to the bottom of the lake. The decomposition of dead material by bacteria is a process which also requires oxygen, thus further diminishing the supply of oxygen in the depth of the lake. This deoxygenation forces species to move to the shallower regions of the lake where oxygen is more plentiful. Here, though, they were more likely to be caught in fishermen's nets or by predators, thus furthering the decline of the cichlid population.

The British started to stock the lake when they realized that fish populations were declining. Officials introduced the perch *Cyprinus carpio* whose introduction does not appear to have been successful and four species of tilapia (*Tilapia zillii*, *T. rendalli, Oreochromis niloticus* and *O. leucostictus*). The only tilapia to take hold so far is a plankton eater, the Nile tilapia (*O. niloticus*), although the other tilapia are present in small numbers. The Nile tilapia is very similar to the endemic species but it can live in a greater variety of habitats, grow at a faster rate, and is a feeding generalist and therefore eats a wider range of foods.

Some British officials also wanted to introduce a large predator such as the Nile perch (*Lates niloticus*) into the lake. Cichlid and some of the other native species are small and somewhat bony. The British felt that a larger, meatier fish would be more pleasant eating, particularly for the restaurant table. Additionally, Europeans enjoyed the perch as a sporting fish because of its large size. The Nile perch, sometimes referred to as the "elephant of the water", can grow to be five or six feet long and weigh up to 135 pounds.

Although most ecologists were opposed to the introduction of the Nile perch because there was no natural predator for it, by 1954 some of this nonendemic species had somehow gotten into the lake. It is possible that floods allowed some of the fish to enter from nearby bodies of water. It is more likely, though, that some perch were intentionally introduced into Lake Victoria. Since it was already there, Ugandan officials decided to complete the process by stocking the lake with Nile perch from Lake Albert.

The Nile perch is a large predatory fish of high commercial and recreational value belonging to the family *latidae*, order *Perciforms*. It is a large mouthed fish, greenish or brownish above, silvery below and generally attains length of 1.8 m and 140 kg. A length of 3 m and a weight of about 200 kg has been recorded. *Lates niloticus* has an elongated, protruding lower jaw, rounded tail and two dorsal fins.

Lates niloticus has two patterns of occurrence, that is endemic and non-endemic. It is endemic in Lakes Albert, Turkana, Chad and Lakes Sharma and Abaya in Ethiopia. In the river systems it is found in the Volta and Nile. In non-endemic environment it occurs in Lakes Kyoga and Victoria, and Kabakas Lake in Kampala. Though it is considered to be non-endemic in Lake Victoria, archeological records indicate that some fossils were found in Rusinga dating back to the Meosine (approximately 25 million years ago). Some fossils have also been found in Lake Edward dating back to Pleistocene period (approximately 35,000 years ago), showing that Nile perch could have been in existence in these Lakes before the recent introduction. It is however not very clear what influenced the present geographical distribution of Nile perch on the continent. *Lates* has not been found in the lakes of Southern part of Africa because of their general depth and their lower temperature as it prefers medium or higher temperatures.

In Lake Victoria, Nile perch was introduced from Lake Mobutu in 1959, 1962 and 1963 around Jinja by officials of Uganda Fisheries Department and some species from Lake Turkana were introduced by officials of Kenya Fisheries Department at Kisumu in 1963. About eight seedlings of perch were introduced at Kisumu point, a number which may have been considered negligible and may not have had any significant biological consequence in Lake Victoria. The main reason for its introduction was to control and convert Haplochromis whose population in the lake was very high into a more desirable and economically viable food crop and extend traditional inshore fishery to offshore waters. Its introduction was also intended to increase the productivity of Lake Victoria as *Lates* grow to a large size, is extremely palatable and is an excellent sport fish. Apart from those reasons it was believed it would do no harm to Tilapia species since it survives with the same in Lake Turkana.

Since its introduction *Lates* has tremendously increased and spread in Lake Victoria. The first catch in the Lake was in May 1960 at Jinja point above Repon Falls and again in November, the same year at the same place. It then began appearing fishermen's catch in large quantities in Uganda waters. By 1965 it had spread round the Northern shore near Entebe and Eastwards to Nyanza gulf and Southernwards to Majita Bay. In the Nyanza gulf, the perch began appearing in the fishermen's catch in the late 1960s and early 1970s. In Kenyan part of Lake Victoria the Nile perch is now well established in the gulf with the highest concentration found on a ridge that runs from Homa point to Uyoma point, areas around Ndere islands in Seme and parts of the gulf proper, mainly in the sandy bays. Kenyan Fisheries Department data recording the trends of fishing in Lake Victoria indicate a steady increase of *Lates* fisheries from near zero in 1968 to about 51 metric tones (0.8%) in 1975 and 27,259 metric tones (about 16%) in 1981. In 1985 the total Nile perch landing was 50,029 metric tones, 51% of total landing. In 1987 the landing increased to 86,833 metric tones corresponding to 69% of the total catches. 1993 estimates from the Kenyan Fisheries department statistical bulletin put the catch of Nile perch to be 99,877 metric tons making it the most abundant and commercially important species in Lake Victoria. The population explosion of this fish in Lake Victoria is attributed to the fact that it predates on other smaller species of fish, can subsist on invertebrates and their brethren, and because of its adaptability to global climatic changes. Other reasons include high fecundity and diversity of habitat as well as the fact that it has a long lifespan. An individual Nile perch may live past 15 years of age.

In general the Nile perch occupies different niches including sandy areas, shallow waters around river inlets and around steep shelving shores, for example Homa Uyoma point in the gulf proper and Namone point in Uganda, which apparently carry a high concentration of oxygen. In the Uganda part of Lake Victoria it occurs mostly within 5 meters of water depth, but has been known to extend sparsely to 20 meters depth. In Lake Turkana perch have been caught at a depth of about 50 meters, but it is reported that the species caught at this level was a smaller one known to be endemic to the lake. In the Nyanza gulf high occurrence of perch is limited to 15 m depth although it has been caught even at depths of 20 meters. The catches of Nile perch decrease with increasing depth from 2.0 kg/hr in the 4-9 m depth zone to 0.4 kg/hr in the depth of 20-29 m zone in Lake Victoria. One researcher caught it exclusively in 10-19 m depth range in the Nyanza gulf with a mean catch of 23.8 kg/hr. The fry of Nile perch are normally restricted to inshore areas within the sublittoral weed beds.

The feeding habits of Nile perch have been studied by several scientists who have concluded that it is a carnivorous fish feeding on dominant species within the lake. However it has been established that Nile perch feeds on a wide variety of different fish species, easily switching to different types and sizes of prey. The major prey of Nile perch when it was introduced in Lake Victoria were detritivorous and planktivorous fishes, but now ecologists believe that it is partly responsible for the disappearance of certain fish species such as *Haplochromine sp.*, *Clarius mosambicus*, *Bagrus docmac* and others in Lakes Victoria and Kyoga. This assertion is supported by the fact that after its introduction, Lake Victoria Haplochromis which formed its main food decreased from 32% of the total catch in 1977 to less than 1% in 1983 and now they are no longer recorded in commercial catches. Likewise since its introduction in Lake Kyoga fish species such as Tilapia species, *Protoptenus aethiopicus*, *Clarius mossambicus*, *Bagrus docmac* and *Haplochromis* have shown progressive decline almost to total depletion during the last two decades of the 20th century. Nile perch feeds on fish predominantly and in addition feeds on prawns, *Caradina nilotica* and dragon fly, Odonata species. *Lates* is also a zooplankton feeder.

Scanty information is available on the reproductive biology of Nile perch. Males mature at a smaller size than the females. Commercial fisheries report peak catches from July to October, but the greatest breeding activity is seen in April just before the long rains. It appears this fish tends not to emigrate far from its territorial grounds for the purpose of spawning. In Lake Turkana the fish that the perch predate on have moved to deeper waters, leaving it on the sandy shallow parts of the lake. The fries which still contain traces of the yolk sac have commonly been found in marginal waters over sandy bottom including aquatic vegetation indicating that these areas are the possible spawning sites. *Lates* shows a very high fecundity of about 15 million eggs per spawning. The eggs hatch into many offsprings which the adults sometimes feed on. The high number of offsprings has led to the well noted high population of Nile perch that has attracted both local and foreign investors.

18.2. The model

The Nile perch is a piscivore and a feeding generalist which means that it eats a wide variety of many different types of food including other fish. This gives it the versatility to adapt to a changing ecosystem and, depending on what stage of its life it is in and on the conditions of the lake, allows it to feed at different trophic levels and within each of these levels on whichever species happens to be abundant. For this reason the logistic equation of Chapter 4 is not a bad approach to try in modeling the Nile perch. If the logistic equation is an accurate model, we would expect the population of perch to stabilize at a high level. However, we could try to take more subtle effects into account. As we saw with the catfish data, there were large apparent oscillations in the population of catfish, ending finally in a steep decline. Might the perch population eventually oscillate too? If so, it would be a result of factors not accounted for in the exponential or logistic models.

Since the main food resource of the Nile perch at first was the cichlids and since the size of the population of cichlids changed drastically between 1971 and 1980, we might expect that this change had an effect on the Nile perch population. We have said that the Nile perch is a feeding generalist, meaning that it eats many different types of prey. This ability has enabled it to survive the drastic decline of the cichlids by eating other species, including its own young. However, if we assume that the perch cannot eat anything but cichlids (which we know is not really the case), then the rate of change of the Nile perch population depends somehow on the number of cichlids available for consumption. In other words, it depends on the amount of available resource. At the same time, the rate of change of the cichlid population depends on the number of Nile perch that are consuming them. It seems as though we will need a system of two differential equations to model the relations between the two populations. In Part 3 on pharmacokinetics we also saw systems of equations in X (the amount of drug in the body), and here we will use a standard notation in which N represents the population of a mid-level consumer (here the cichlids) and P represents that of the predator (here the Nile perch).

An equation that describes the cichlid population will have to take into account the rate of change of the population if there were no predator present (a positive rate of change) and the rate of change of the prey population due to predation (a negative rate of change). Very generally for the cichlid population we have:

$$\frac{dN}{dt} = \text{(rate of population increase)} - \text{(rate of population decrease)}.$$

If for simplicity we use the Malthusian expression for exponential population increase then where r is a positive constant:

$$\frac{dN}{dt} = rN - p(N, P).$$

The function $p(N, P)$ describes the rate of predation as it varies with changes in the perch and cichlid populations.

A simple version of $p(N, P)$ may assume that the rate of predation is related to the frequency with which a perch and a cichlid come into contact with one another. If we assume that the members of each population move randomly and are evenly distributed, by multiplying the cichlid and perch population densities we can approximate the frequency of their encounters. By multiplying this NP term by a positive constant (b) we can represent the number of encounters that result in prey death. By then subtracting this from the rate of cichlid population increase expected if growth were unlimited (rN) we arrive at the equation:

$$\frac{dN}{dt} = rN - bNP.$$

Similarly, the rate of change of the size of the perch population is dependent on its rate of predation because the consumption of cichlids adds to the perch population. The population of perch also depends on the rate of decrease of the population without any prey to eat. Therefore:

$$\frac{dP}{dt} = \text{(rate of increase)} - \text{(rate of decrease)}.$$

If there were no cichlids, we could assume that the perch would die at a constant rate because there is no food, and that it therefore would have a death rate that follows a pattern of negative exponential growth $(-gP)$. Like the bNP term in the cichlid equation, we would want to include a term in the differential equation for the perch population that describes the changes in the perch population as the cichlid population changes. Just as the cichlid population lost a certain number of its members, the predator population will gain a number of members proportional to the number of encounters (NP) between the two species. We will use another constant (c) which is the number of predator births that result from each encounter. The resulting differential equation describing the change in perch population is:

$$\frac{dP}{dt} = cNP - gP.$$

EXERCISE 18.1. The term NP in both differential equations encodes certain assumptions about the two populations and how they might encounter each other. Describe some of them in your own words.

The set of equations describing the predator-prey relationships, rearranged for more convenient analysis is:

$$\frac{dN}{dt} = rN - bNP$$

$$\frac{dP}{dt} = -gP + cNP.$$

The ratio of the constants b and c reflects the relative ease of converting prey into predator, in this case cichlid into perch. A predator like the Nile perch may be commercially more profitable than the smaller fish it eats, but it is higher in the food chain. Because energy stored in its prey is lost in the process of hunting, catching and metabolising prey, it is a less efficient "crop" than an organism lower in the food chain such as the cichlid. No consumer is able to convert all of the energy bound up in its prey into energy it can use. Biologists estimate that at each trophic level 80 percent of the energy stored in the previous level is lost from the food chain. Put another way, one kilogram of Nile perch produced means a corresponding loss of four kilograms of other fish. This estimate of the rate of energy transfer is incorporated into the equations by the constants, b and c.

18.3. Analysis of the model

The conditions under which neither the prey population nor the predator population size is changing (the birth rate equals the death rate for each population) is called the equilibrium of the system. To find the equilibrium we set $\frac{dN}{dt}$ and $\frac{dP}{dt}$ both equal to zero:

$$\frac{dN}{dt} = 0 = N(r - bP)$$

$$\frac{dP}{dt} = 0 = P(-g + cN).$$

One equilibrium solution to the cichlid equation is when both $N = 0$ and $P = 0$. Another equilibrium solution occurs when some perch and cichlids are present (neither N nor P is zero). In this case

$$r - bP = 0 \text{ and } -g + cN = 0.$$

Solving these equations we find that $P = r/b$ and $N = g/c$.

For each of these equilibrium solutions, the size of each population is constant. In real ecosystems population size does sometimes fluctuate. What if the size of either or both of the populations changed a little from its equilibrium value? Would the system go back to the equilibrium state or would the size of the populations change drastically? In order to find the answers to these questions we need to determine whether the equilibrium solutions are stable.

The phase plane that we will use to analyze the equilibria of the predator-prey system has N, the size of the cichlid population, as its horizontal axis variable, and has P, the size of the perch population, as its vertical axis variable. This type of phase graph is a combination of phase graphs for the perch and the cichlid equations and is shown in Figure 18.1. We leave out the t axis because the values of N and P at a given time are completely determined by our their starting values. Once again the equilibria of the system appear as points on the phase graph, $(0,0)$ and $(g/c, r/b)$. If we choose the constants $r = 1$, $b = 0.5$, $c = 0.5$, and $g = 1$ for our phase graph, the equilibria would be at $(0,0)$ and $(2,2)$.

EXERCISE 18.2. In Figure 18.1, the equilibrium $(0,0)$ can be identified visually. By computing the vectors in the phase portrait in Figure 18.1 at $(0, \pm 0.1)$ and $(\pm 0.1, 0)$, conclude that the equilibrium point $(0,0)$ seems to be unstable. Conclude that the populations can get arbitrarily close to 0 and still recover.

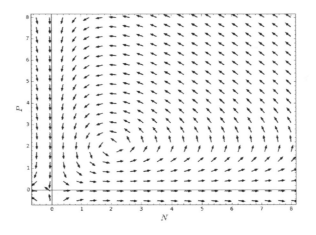

FIGURE 18.1. Phase portrait for predator-prey equations, Lotka-Volterra model. In this case we plotted

$$\frac{d\mathbf{X}}{dt} = \begin{pmatrix} (0.5N - 1)P \\ (1 - 0.5P)N \end{pmatrix},$$

where we consider the system in vector notation. The equilibria of $(0,0)$ and $(2,2)$ can be identified in the graph. The point $(0,0)$ is an equilibrium because if you put a particle down at that point, it will not flow anywhere since it has no vertical motion (the arrows point in opposite directions just above and below the point) and it has no horizontal motion (the arrows point in opposite directions just to the left and right of the point). The point $(2,2)$ is an equilibrium because if you put a particle down at that point, it will just spin as the phase portrait flows around that point.

EXERCISE 18.3. In Figure 18.1, the equilibrium $(2,2)$ can be identified visually. By computing the vectors in the phase portrait in Figure 18.1 at $(2 \pm 0.01, 2)$ and $(2, 2 \pm 0.01)$, conclude that the equilibrium point $(2,2)$ seems to be stable because the populations are swirling around $(2,2)$.

Now we may use the differential equations of the system to determine the approximate paths of trajectories that start near one of the equilibrium points. A trajectory that begins near the equilibrium point $(0,0)$ could start with $N = 0$ and $P = a$, where a is a small positive quantity. This means that there are no cichlids but there are some Nile perch. We would expect that since the perch have nothing to eat, they would die off: $\frac{dN}{dt}$ would stay at zero and N would remain at zero because cichlids cannot be born without parents. Thus the perch would die off exponentially as a result of our original assumptions. It appears that $(0,0)$ is stable if we look only at this one starting point. See Figure 18.2. The trajectory starting at $(0, a)$ lies on the P axis.

If instead we had started with the conditions $P = 0$ and $N = a$, we would expect exponential growth of the cichlid population. One of our main assumptions about the system is that the prey population, without any predator, will grow exponentially. Analysis of the equations at $(a, 0)$ shows this to be the case. $\frac{dN}{dt} =$

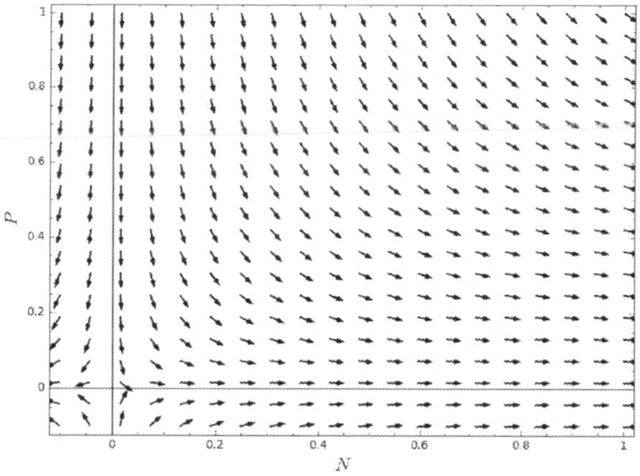

FIGURE 18.2. A close-up view of Figure 18.1 around the origin. Observe that at $N = 0, P = a$ where a is small, the equations

$$\frac{dN}{dt} = N(r - bP)$$

$$\frac{dP}{dt} = P(-g + cN)$$

suggest that the arrow attached to $(0, a)$ should be vertical and pointing downward, an observation that is consistent with the graph and with the biological fact that if there are no cichlids, the nile perch should die off. Also observe that at $P = 0$ and $N = a$, where a is small, we see that N keeps getting bigger and bigger, consistent with the biological fact that in the absence of a predator we have assumed that the cichlids would enjoy exponential growth.

rN. It appears that $(0, 0)$ is unstable if we look only at this one starting point, as can be seen in Figure 18.2.

If both the perch and the cichlid populations start out with small positive values ($N < g/c$ and $P < r/b$), $\frac{dN}{dt}$ would be positive and $\frac{dP}{dt}$ would be negative. The perch would not have enough to eat and so it would die off. At the same time, there would not be enough perch to prey on all of the cichlids. The cichlid population would grow, but at a slower rate than it would without the presence of any Nile perch.

The fact that most solutions tend away from $(0, 0)$ but at least one tends towards it shows us that this unstable equilibrium is something mathematicians call a saddle point.

EXERCISE 18.4. Summarize what you just read, by explaining in your own words why $(0, 0)$ is a saddle point.

If we test small divergences from $(g/c, r/b)$ we get the trajectories shown in Figure 18.3. This equilibrium point seems to be a center point because the trajectories appear to form closed loops around it. The computer program used to generate this graph starts with one value of N and one value of P and then plots the approximate path of the trajectory by using $\frac{dN}{dt}$ and $\frac{dP}{dt}$ to estimate the changes

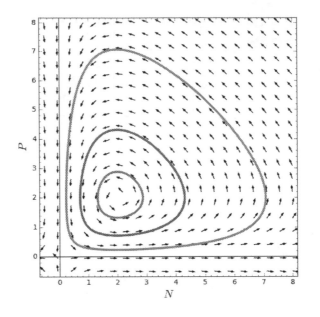

FIGURE 18.3. Typical trajectories for solution to Lotka-Volterra: Perch versus cichlid.

in N and P that give us the next point. What direction are the trajectories traced over time? We can use the differential equations to find the direction of change of N and of P. If we start with $P < r/b$ and $N < g/c$, $\frac{dN}{dt}$ is positive, and N increases. However, $\frac{dP}{dt}$ is negative and P decreases. If $P < r/b$ but $N > g/c$, both N and P increase because both $\frac{dN}{dt}$ and $\frac{dP}{dt}$ are positive. If $N > g/c$ and $P > r/b$, N increases and P decreases. Finally, if $N < g/c$ and $P > r/b$, both N and P decrease. These qualitative changes show us that the trajectories are traced in a counterclockwise direction around the equilibrium point. The more sophisticated approach the computer uses indicate that the solutions oscillate around the equilibrium but they never reach it.

EXERCISE 18.5. Summarize what you just read, by explaining in your own words why $(2, 2)$ is a stable equilibrium around which the populations oscillate.

The cichlid and perch population sizes will change between values above and below those of their equilibrium. Each will pass through its equilibrium value, but never both at the same time, so the system will never be at equilibrium. We can see these oscillations if we plot the Tanzania catch data for Nile perch and cichlids against time, as shown in Figure 18.4. A phase graph of the data is shown in Figure 18.5.

Figures 18.4 and 18.5 illustrate the utility of the phase portrait. If we only looked at graphs of the perch with cichlid populations over time, we might conclude that they oscillate more or less as predicted by the Lotka-Volterra (predator prey) equations we have been studying. Looking at Figure 18.5, however, tells us that our model is very far from what the data are doing. We already knew the perch ate more than one prey fish, but if we didn't know that, then comparing the phase

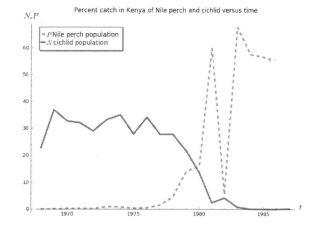

FIGURE 18.4. Percent catch in Kenya of Nile perch versus time and percent catch in Kenya cichlid versus time based on data found in [7] and [6]. The dashed is the Nile perch population and the solid line is the cichlid population. Compare this with Figure 18.6.

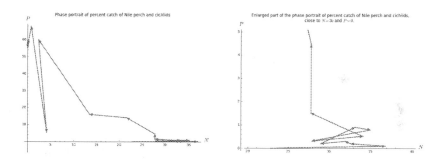

FIGURE 18.5. The figure on the left is a phase portrait of percent catch in Kenya of Nile perch versus percent catch in Kenya of cichlid and the figure on the right is the result of zooming in near the point $(30, 0)$. The arrows present the change in population from one year to the next. Compare this figure to Figures 18.1 and 18.3.

graph of our model against the data would be a big clue that we had overlooked some factors.

Figure 18.6 shows solutions to the predator - prey plotted over time. As you see, it is harder to see from this kind of graph that something is drastically wrong with our model. Although the model captures the ups and downs of fish populations, it does so in a periodic way that is not particularly close to what the data show. As with the logistic equation, we have been able to model a general qualitative feature of the system but with poor ability to predict specifics. For fisheries catch data, this failure is not surprising, and this is probably the right place to talk generally about fish models.

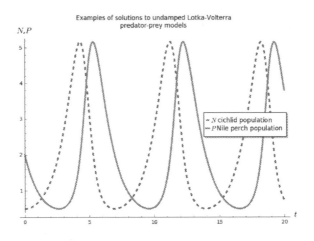

FIGURE 18.6. A plot of predator (the solid line, say) and prey (the dashed line, say) versus time. They are the plots of the particular solutions to

$$\frac{dN}{dt} = N(r - bP)$$
$$\frac{dP}{dt} = P(-g + cN)$$

with $g = r = 1$, $b = c = 0.5$ and $P(0) = 2$ and $N(0) = 0.5$.

18.4. Implications of the model

During its life cycle a typical fish may play the role of prey, predator, or both. Many species of fish lay eggs, which are prey for a large variety of organisms that may themselves be prey for hatchlings, fingerlings or older versions of the same fish. Some fish, such as the Nile perch, are such feeding generalists that they routinely consume many of their own offspring. So a more complete model for the Nile perch would include boxes for different sizes of perch, with species interactions specific to those sizes. Furthermore, one predator left conspicuously out of the model in this chapter is human. Catch data, in particular, is not just a benign measure of the fish population but the result of large scale predation on perch by human beings. Our predation is related to the size of the fish population, but is also controlled by a variety of regulations concerning seasonal control, daily catch limits, mesh size of nets and other factors. Further, in Lake Victoria both the perch and their prey are fished, but in different locations, with different methods. The human predator has different behavior with respect to different species. A more useful model would take some of these factors into account. What follows in this chapter is a brief description of the fishing industry in Lake Victoria at the end of the 20th century.

The effect of the introduction of *Lates* on the ecology of the lake was matched by the effect of improved fishing techniques in the region. Together these changes had enormous effect on the local economy. Larger catches, sophisticated equipment, processing plants and a commitment to export drastically changed the way everyone in the region made their living.

The fishing gears drastically changed from simple traps to sophisticated indiscriminate fishing gears that included the use of trawl nets, complicated gill nets, long hook lines with better ways of lowering fish and wide beach seining facilities. The fish that were fully in demand by this time included Tilapia species, catfish, (*Curius mossambicus*) and lungfish because of its tasty firm meat. However with the introduction of dairy, poultry and fish farming, *Rustreoneobola argentia* (omena) which is massively used as animal feed soon became the focus for commercial beach seining. The seine for omena is intended to catch small sized fish and hence could not discriminate the small haplochromis and juveniles of other fish species. It is likely that the continual use of this type of gear could have contributed to the disappearance of most of the fish species because it indiscriminately catches the fingerlings which are crucial for the procreation of species. In the early 1970's as much as 35 percent of the commercial catch was immature. Researchers concluded that the larger haplochromis had almost been eliminated by overfishing. The omena (sardine) industry thrived, thereby encouraging processing plants. The processing of this fish for animal feed involves the drying of the fish for several days followed by grinding and packaging. Omena is also a rich source of proteins and has been used in hospitals for protein replenishment of malnourished children. Since it was being caught in tons it became popular with the public as the cheapest source of fish protein. Rich individuals invested in fisheries and sooner or later the catch had to dwindle with or without the introduction of other factors into the ecosystem.

Although pollution and introduction of Nile perch might be blamed for the disappearance of several species, the rate of overfishing and indiscriminate fishing without corresponding stocking of the lake has quite possibly contributed more to the disappearance of several Lake Victoria species. With the systematic decrease in the population of other fish species, the population of Nile perch dramatically increased. The rapid increase in the population of this fish has been attributed to its high fecundity and the availability of varieties of food within the lake. The population explosion of this fish brought in an alternative for fish processing industries whose survival were now threatened by low catch of omena.

Several Nile perch processing plants have mushroomed around lake Victoria with high concentration on the Kenyan side. The processing of Nile perch starts immediately when it is caught. Due to high temperatures the boats must be equipped with cooling facilities to reduce spoilage during long fishing episodes. The fishermen who lack these facilities are heavily exploited by middlemen who provide cooling facilities in exchange for low prices. Several cooperative societies established with the intention of saving the fishermen from this kind of exploitation have failed. The fishermen therefore are at the mercy of the middlemen and the large processing plant owners. The roads to the Nile perch collection points are poor and pathetic. The refrigerated trucks get stuck during the rainy season, leading to spoilage of fish before they reach the processing centers. There is no electricity supply, treated water and telephone connection to these centers. It is therefore unfortunate to realize that the exploitation of the Nile perch by these processing plants does not necessarily improve the economic status of the local community.

Chilled whole Nile perch are landed on the beach where they are weighed and transported to the processing plants which are located in the major cities, especially Nairobi and Kisumu in Kenya. Kisumu has the highest concentration of fish processing plants since it is located on the shores of Lake Victoria. The processing of Nile perch in the factory starts by filleting, and deskinning. The fillets are taken to the trimming section where they are into 2 kg size for easy packaging. They are then packaged and stored into huge coldrooms awaiting exportation. The fillets processed this way are very expensive and are exported to Israel and the Middle East, the United Kingdom, and other western countries. The factory owners at one point paid approximately $0.1 per kg for unprocessed Nile perch fillet but sold it at about $5 or more per kg, thereby making a large profit. Processing factories mainly utilize the flesh as fillets for source of proteins, although no part of the Nile perch is wasted. The skin is used for making leather, the air sack for making surgical threads, the fats from the abdominal cavity for cooking oil and the carcass (mgongo wazi) sold back to the small scale business women who fry it and sell it locally. The local community which participates in the conservation of the lake is finally given the carcass simply because it cannot afford the fillet.

The processing plants are complex facilities employing large numbers of workers. Most often they are unskilled laborers working for a day's wage. They are generally paid low salaries because this region is overpopulated with a high rate of unemployment. Most of the workers are therefore hired on temporary basis and are not entitled to any benefits and do not belong to any union which can defend their rights. This situation is not easy to correct because the fish processing plants are run by rich individuals who capitalize on their connections to the high powers in the government system and on the ignorance of their workforce. Due to an unskilled inconsistent laborforce and undefined divisions of the compartments of the processing stages, there have been several reports of cross contamination of the finished products with such deadly bacteria such as *E. coli* and salmonella. A lot of consignments have been rejected and destroyed in the countries of destination. Because of this, the UNESCO Food and Agriculture Organization (FAO) installed a set of hygienic conditions that must be met by the processing plants to be allowed to export. These conditions included employment of regular trained individuals occasionally checked by medical personnel, reorganization of the plant so that the offloading zone for whole fishes is separate from the finished products, presence of treated running water and sanitized floor and other facilities. The factory should also be inspected by Fishery officers at least twice a month. Plants that meet these requirements are allowed to use specific stamps on their finished products to allow them to export to European countries. However, the problem is far from over because most of the fish are exported to Middle Eastern countries which are not so concerned with these conditions.

The export of Nile perch has recently become one of the major foreign exchange earners for Kenya. However with the high rate of exploitation there is a likelihood that the Nile perch business might soon be faced with eminent extinction. There are several ways by which this industry can be maintained. One suggestion would be to create certain fish breeding zones within the lake where absolutely no fishing is conducted. The other alternative would be to continually restock the lake with not only Nile perch but also with other fish species which are rapidly disappearing. Since the Nile perch business generates a lot of income, encouraging aquaculture is also a possibility. However this will require large areas or large ponds, since this fish grows into very large sizes.

For more information on Nile Perch, and modeling their population, see [2], which includes the constants used for Nile Perch models; see [4] for Nile Perch models with an explanation of political ramifications and see [5] for data. For a broader base of understanding of Nile Perch and other introduced fish in Lake Victoria, without modeling, [3] provides a reference. For a better understanding of fish population changes in lakes other than Lake Victoria, and with other fish, [1] and [8] give models with relevant parameters which can bring insight to modeling Nile Perch, or understanding the mathematics of fish in general.

18.5. Plotting phase portraits in Sage

In this chapter, we have seen a number of phase portraits for predator prey models. To plot a figure such as Figure 18.3, you need to plot the slope field and then the trajectories. We break the process into two steps.

First, recall that we want to make a phase portrait for the system

$$\frac{dN}{dt} = N\left(1 - \frac{P}{2}\right)$$
$$\frac{dP}{dt} = P\left(\frac{N}{2} - 1\right);$$

and maybe plot trajectories through particular points.

Let's step back and think about this a little more generally. A phase portrait attaches to every point (N_0, P_0) in the phase plane a vector whose Nth-component (the Nth-component is, algebraically the first component and geometrically it is the horizontal component) is $\frac{dN}{dt}$ at the point (N_0, P_0) and whose P component is $\frac{dP}{dt}$. In this system, for example, at the point $(5, 7)$ we would attach the vector

$$\begin{pmatrix} N_0\left(1 - \frac{P_0}{2}\right) \\ P_0\left(\frac{N_0}{2} - 1\right) \end{pmatrix} = \begin{pmatrix} 5\left(1 - \frac{7}{2}\right) \\ 7\left(\frac{5}{2} - 1\right) \end{pmatrix} = \begin{pmatrix} -\frac{25}{2} \\ \frac{21}{2} \end{pmatrix}.$$

We would continue to do this with a sampling of points if we were doing it by hand.

In Sage, the code

```
f(N,P) = ((-.5*P+1)*N,(.5*N-1)*P)
v=plot_vector_field(f,(N,-0.5,8),(P,-0.5,8),headaxislength=5,
    headlength=4,axes_labels=['$N$','$P$'])
show(v)
```

makes a plot like Figure 18.7. The first line of this code snippet says to make a two-variable function whose output is a (2-dimensional) vector (such a function is called a *vector field*). The next line says to plot the vector field f for N ranging from -0.5 to 8 and for P ranging from -0.5 to 8. The last three parameters have to deal with the shape of the arrow and the labels on the axes.

While Figure 18.7 is an accurate rendering of the phase portrait for the system, you might notice that near the point $(2, 2)$ the arrow is almost invisible. This is not surprising because if you calculate what arrow should be attached to the point $(2.1, 1.9)$, say, you get

$$\begin{pmatrix} P_0(0.5N_0 - 1) \\ N_0(1 - 0.5P_0) \end{pmatrix} = \begin{pmatrix} 1.9(0.5 \cdot 2.1 - 1) \\ 2.1(1 - 0.5 \cdot 1.9) \end{pmatrix} = \begin{pmatrix} 0.095 \\ 0.105 \end{pmatrix}$$

which is very small. So, in Figure 18.8 we plot a related vector field, one that is *normalized* so that all vectors have the same length. A vector

$$\mathbf{w} = \begin{pmatrix} 1 \\ 4 \end{pmatrix}$$

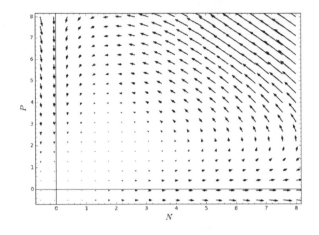

FIGURE 18.7. Phase portrait for the system $dN/dt = N\left(1 - \frac{P}{2}\right)$ and $dP/dt = P\left(\frac{N}{2} - 1\right)$. Note that at the point $(5, 7)$ we attach a vector whose horizontal component is negative and whose vertical component is positive and that this is consistent with the calculation in the text. The size of the components are not those that we calculated in the text because Sage makes some choices about the lengths of the vectors so that the resulting figures are easier to digest by the reader.

can be thought of as an arrow from the point $(0,0)$ to the point $(1,4)$. Its length, then, should be, by the Pythagorean theorem, $\sqrt{1^2 + 4^2} = \sqrt{17}$. In general, an arbitrary vector

$$\mathbf{v} = \begin{pmatrix} x \\ y \end{pmatrix}$$

has length $\sqrt{x^2 + y^2}$ and is denoted $||v||$. To construct a new vector $\hat{v}$ whose direction is the same as $\mathbf{v}$ but whose magnitude is 1, we simply scale $\mathbf{v}$ by its length. E.g., the vector

$$\begin{pmatrix} \frac{1}{\sqrt{17}} \\ \frac{4}{\sqrt{17}} \end{pmatrix}$$

is in the same direction as $\mathbf{w}$ but has length 1.

EXERCISE 18.6. Verify the previous claims about the vector

$$\begin{pmatrix} \frac{1}{\sqrt{17}} \\ \frac{4}{\sqrt{17}} \end{pmatrix}.$$

Namely, that it is in the same direction as $\mathbf{w}$ and has length 1.

So, to make a plot like the one in Figure 18.8, we write

```
g(N,P) = ((-.5*P+1)*N/sqrt((-.5*P+1)^2*N^2+(.5*N-1)^2*P^2),
          (.5*N-1)*P/sqrt((-.5*P+1)^2*N^2+(.5*N-1)^2*P^2))

v = plot_vector_field(g,(N,-0.5,8),(P,-0.5,8),headaxislength=5,
      headlength=4,axes_labels=['$N$','$P$'])
show(v)
```

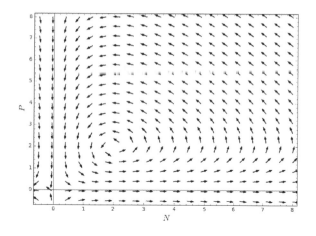

FIGURE 18.8. Phase portrait for the system $dN/dt = N\left(1 - \frac{P}{2}\right)$ and $dP/dt = P\left(\frac{N}{2} - 1\right)$. We have normalized the vectors in the plot so that they all have the same length.

where the only difference between this code snippet and the previous one, is that we have divided each component of the vector field f by its length to get this new vector field g. We find the normalized phase portrait easier to understand.

Now, to plot trajectories of particular solutions on the phase portrait, we need to solve the differential equation and plot the solutions. In particular, say that we want to solve the IVP

$$\frac{dN}{dt} = N\left(1 - \frac{P}{2}\right)$$
$$\frac{dP}{dt} = P\left(\frac{N}{2} - 1\right)$$
$$N(0) = 1, P(0) = 4.$$

In Sage, we solve the system as usual and put it altogether as

```
f(N,P) = ((-.5*P+1)*N/sqrt((-.5*P+1)^2*N^2+(.5*N-1)^2*P^2),
          (.5*N-1)*P/sqrt((-.5*P+1)^2*N^2+(.5*N-1)^2*P^2))

v = plot_vector_field(f,(N,-0.5,8),(P,-0.5,8),headaxislength=5,
        headlength=4,axes_labels=['$N$','$P$'])

N,P,t=var('N p t')

ss = desolve_system_rk4([(1-P/2)*N, (N/2-1)*P], [N,P],ics=[0,1,4],
        ivar=t,end_points=100)
Q = [(NN,PP) for tt,NN,PP in ss]
v += list_plot(Q,plotjoined=True,axes_labels=['$N$','$P$'],
        thickness=3)

show(v)
```

We observe that the trajectory is plotted at the end. The list Q of points is a collection of points where the first coordinate is a value of N and the second is

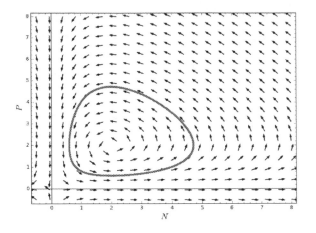

FIGURE 18.9. Phase portrait for the system $dN/dt = N\left(1 - \frac{P}{2}\right)$ and $dP/dt = P\left(\frac{N}{2} - 1\right)$ with a tracjectory through the point $(1, 4)$.

a value of P: recall that this is the case because the differential equation solve procudes a list ss of points (t, N, P) and in the construction of Q we are just grabbing the second and third components of Q.

18.6. Problems

PROBLEM 18.7. The relationship between human and Nile perch is also one of predator to prey. Model this relationship using a pair of equations describing change in each of the two populations. What are your assumptions about fish and human?

PROBLEM 18.8. How does the nature of the model change depending on whether the exponential or logistic equation is the basis of the model? Which population (predator or prey) does it make sense to cap with a logistic term? Use a computer to draw phase portraits of your system and investigate the long term effects of these two assumptions.

PROBLEM 18.9. How would you modify your model to include the lower level prey (haplochromis, or cichlid)? What if a major portion of the adult perch's diet is juvenile perch? What if cichlids and other small fish also eat perch eggs? There is a web of relationships here that can be sorted out with assistance from the literature.

PROBLEM 18.10. One of the most challenging aspects of ecological models is determining the constants of growth, predation, energy transfer up the trophic level, etc. Pick a variation on the model in this chapter and go to the literature to find these for a specific situation.

PROBLEM 18.11. How would you adjust your human-perch model to account for the restocking of the lake with perch, as is suggested by various people? How would you account for the designation of no-fishing areas, as is also suggested? Be sure to state your assumptions about fish behavior carefully and take them into account in your equation.

Bibliography

[1] Donald L DeAngelis, William F Loftus, Joel C Trexler, and Robert E Ulano-
    wicz. Modeling fish dynamics and effects of stress in a hydrologically pulsed
    ecosystem. *Journal of Aquatic Ecosystem Stress and Recovery*, 6(1):1–13, 1997.

[2] JYT Mugisha and H Ddumba. Modelling the effect of Nile perch predation and
    harvesting on fisheries dynamics of Lake Victoria. *African Journal of Ecology*,
    45(2):149–155, 2007.

[3] M Njiru, E Waithaka, M Muchiri, M Van Knaap, and IG Cowx. Exotic intro-
    ductions to the fishery of Lake Victoria: What are the management options?
    Lakes & Reservoirs: Research & Management, 10(3):147–155, 2005.

[4] Richard Ogutu-Ohwayo. The decline of the native fishes of lakes Victoria and
    Kyoga (East Africa) and the impact of introduced species, especially the Nile
    perch, *Lates niloticus*, and the Nile tilapia, *Oreochromis niloticus*. *Environmen-
    tal biology of fishes*, 27(2):81–96, 1990.

[5] Richard Ogutu-Ohwayo. Management of the Nile perch, *Lates niloticus* fishery
    in Lake Victoria in light of the changes in its life history characteristics. *African
    Journal of Ecology*, 42(4):306–314, 2004.

[6] FL Orach-Meza. Present Status of the Uganda Sector of Lake Victoria Fisheries,
    1992.

[7] J Eric Reynolds and DF Greboval. *Socio-economic effects of the evolution of
    Nile perch fisheries in Lake Victoria: a review*. Number 17. Food & Agriculture
    Org., 1988.

[8] Lin Wu and David A Culver. Daphnia population dynamics in western Lake
    Erie: regulation by food limitation and yellow perch predation. *Journal of Great
    Lakes Research*, 20(3):537–545, 1994.

CHAPTER 19

Damped Lotka-Volterra Equations

The poor match between catch data and the predictions of the predator-prey model of the last chapter is a result of multiple factors. In particular there are three difficulties worth thinking about with regard to that model.

First of all we have to question the reliability of catch data as a reflection of the fish populations. As the chapter pointed out, during the years that data was collected, fishing underwent major changes in the region. A broader range of fish were caught and more resources were put into catching the larger perch. Fish populations may not have been quite as erratic as the data suggest and some of the randomness may be the result of changes in fishing rather than fish.

Second, the environment and therefore the species living in it are subject to random effects of weather, human activity and other forces. Only in controlled laboratory conditions do we tend to see models conforming closely to actual data. The real world is messy, with random shocks to an ecosystem being the norm rather than the exception. Of course it is possible to model random changes. One could build a program that randomly enlarges or diminishes a population by a small amount at every time step of the numerical algorithm. Done correctly, this would result in an output that looked a bit more like real data in general but not more like any particular data set.

Third, we may have made a model that failed to incorporate some important assumptions about the situation. This is certainly true of the model in the last chapter. The cichlid is not one single species, but many. It competes for resources with other small fish, none of which are in the model. The Nile perch is a feeding generalist and eats those other fish, so it is not solely reliant on cichlid, unlike what our model would have us assume. The Nile perch even eats its own juveniles, functioning at times as both predator and prey. So there are lots of aspects of the real situation that are not accounted for in the model.

19.1. Biological context

The utility of a model does not always lie in its ability to match a data set, especially one heavily influenced by more or less random events. We could always try to fit a function to a set of data points and there are many ways to do so. If we were to fit a polynomial to the data set in the last chapter, it might fit perfectly. In the long run, though, it would behave like a polynomial, going up infinitely far or else becoming negative and dropping. Alternatively, we could fit a sum of sines and cosines of different amplitudes and frequencies. If we used enough of them, we could fit the data exactly. In the long run, our function would oscillate endlessly, because of the choice of functions we used to build it. Then we would have to decide whether we believe the polynomial answer that, say, rises indefinitely or whether we believe the oscillating answer. Our beliefs are, of course, not a reliable method of predicting the future. If they were, we wouldn't bother to make a model at all.

19.2. The model

The point of modeling with differential equations is that we build our assumptions about how nature works into the equations themselves as hypotheses. The solutions then tell us which macroscopic features will follow from those hypotheses. In the last chapter one hypothesis was that growth in predator population depended on quantity of prey available, according to a rule that we prescribed. The first part of this hypothesis is not a huge assumption but the actual growth rule we used could certainly be adjusted. We also assumed that the prey was diminished in a way that depended on the number of predators. Also perfectly reasonable, except that the actual rule we prescribed might be improved. Finally we made assumptions about what these two populations would do in the absence of the other one. If there is no prey, the predator follows the rule:

$$P' = -gP.$$

The predator would thus die out exponentially. If there is no predator, then the prey follows the rule:

$$N' = rN.$$

The prey would thus grow exponentially. Together the full set of equations was:

$$N' = rN - bPN$$
$$P' = -gP + cNP.$$

So these are our hypotheses. The long term behavior predicted by the model has two main features. First of all, there is an equilibrium solution. Second, all of the other solutions are periodic, oscillating forever. In fact, oscillations of all sizes are possible, with populations of either predator or prey getting arbitrarily large.

The predator-prey equations cleanly demonstrate the difference between modeling with differential equations versus just fitting functions to data. In the differential equation model we make assumptions about how nature works now, and see what the model predicts in the future. To fit a curve to data we don't need to make any assumptions, but if we want to know what will happen in the future we have to assume we know what the future will be in advance of choosing which functions to use to approximate our data. The first kind of modeling is more powerful because our assumptions are about what is happening in nature now, therefore they can be tested, observed, and debated more usefully than our guesses about the future.

Of all the hypotheses we built into these equations, the one easiest to argue about is that the prey population is capable of limitless growth. Nowhere in nature do we observe a species to have this property. Other than economists discussing the stock market, nobody ever believes limitless growth of anything to be a reasonable prediction for the future. But in the lab, we do sometimes see growth following a logistic curve. So we could argue pretty convincingly that, even though it won't be perfect, a model that results in logistic growth for prey in the absence of predator would be a better hypothetical situation than what we have used so far. To do this we need only change one term:

(19.1) $$N' = rN(1 - N/K) - bPN$$

(19.2) $$P' = -gP + cNP.$$

Now if there is no predator, N follows the rule:

$$N' = rN(1 - N/K).$$

This is the logistic equation from Chapter 4. The growth of N is limited by the number K, the carrying capacity of its habitat. For the purposes of comparing models, we assume $K = 1$. Think of the units of N as being "percent of carrying capacity". Here is one solution typical of this model, first displayed as a time series and then as a phase portrait. The time series for the solution to this system can be seen in Figure 19.1 and the phase portrait can be seen in Figure 19.2.

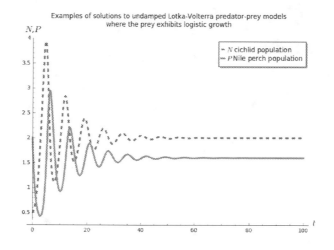

FIGURE 19.1. Time series for the solutions to (19.1) with $r = 1$, $K = 10$, $b = -0.5$, $g = -1$ and $b = 0.5$ and with initial conditions $N(0) = 0.5$ and $P(0) = 2$.

19.3. Analysis of the model

An important qualitative difference has resulted from changing just one assumption. Whereas our original predator/prey equations had solutions that oscillated forever with no reduction in amplitude, these solutions show damped oscillation tending eventually to equilibrium. Comparing these two models supports the assertion that limits to growth are a factor that causes populations in a predator/prey relationship to approach equilibrium.

The tendency of populations to reach an equilibrium value and stay there is the subject of much discussion among ecologists and modelers. Often we assume that an ecosystem was in equilibrium until we humans came along and disturbed it. There usually are no data to support such an assumption; just the general belief that, left alone, a system would behave thus. Reasoning completely backwards (which we should never do) we would prefer the damped predator/prey equations because they produce an answer we would like to believe. Fortunately we do not have to reason backwards because the assumption of limits to growth is far easier

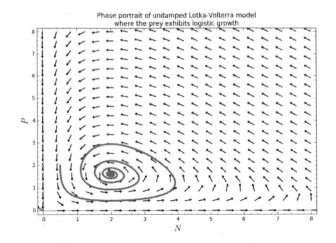

FIGURE 19.2. Phase portrait for the solutions to (19.1) with $r = 1$, $K = 10$, $b = -0.5$, $g = -1$ and $b = 0.5$ and with initial conditions $N(0) = 0.5$ and $P(0) = 2$.

to justify and, in this case, it produces solutions tending to a limiting equilibrium, a property that we desire.

19.3.1. Perturbations. Returning to the discussion of random events at the start of this chapter, we might investigate how the two systems we have been studying would respond to such a perturbation. In Figure 19.5 we see phase portrait illustrating the effect of perturbing the initial conditions for the undamped Lotka-Volterra model considered in the previous chapter and in Figure 19.4, we see the effect on the time series.

If an event (such as fishing) suddenly lowered the population of predator, one of two qualitatively different things would happen. Figure 19.5 shows two possible points where a lowering of predator would have different effects. If the predator were removed at point A, the solution moves to a trajectory outside its original one. It moves farther from the equilibrium value and it oscillates with greater amplitude indefinitely. If the predator is removed at point B, the solution moves inside its original trajectory and oscillations are reduced in amplitude. Note that at A and B the prey population is exactly the same.

Notice that in Figures 19.5 and 19.4, the perturbation is propagated throughout time. That is, a single removal of a small amount of one population creates a change in amplitude of the oscillation that is propagated forever.

By contrast, the damped predator prey equation, whose phase portraits and time series are seen in Figure 19.6 the phase portrait looks like Figure 19.5, with trajectories that spiral into a fixed point. Once again, we could reduce the predator at one of two locations. But because all solutions spiral in toward the equilibrium solution, eventually traces of the random event will disappear, as shown in the time series plots in Figure 19.5.

For these models, the introduction of an assumption of limited growth is actually a mitigating factor against random perturbations. In both cases an ill-timed reduction of predator population can result in a rebound effect where the predator

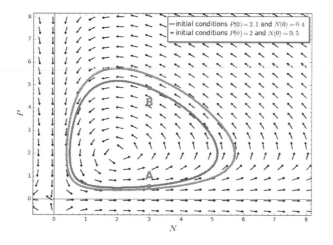

FIGURE 19.3. Two trajectories for the undamped Lotka-Voletta system (19.1) with $r = 1$, $b = -0.5$, $g = -1$, $b - 0.5$, and where the prey exhibits intrinsic exponential growth. The inner trajectory corresponds to initial conditions $N(0) = 0.5$ and $P(0) = 2$ and the outer trajectory corresponds to initial conditions $N(0) = 0.4$ and $P(0) = 2.1$.

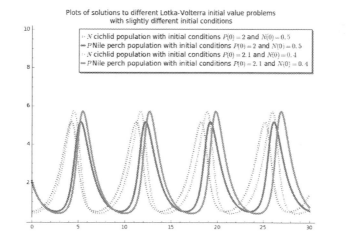

FIGURE 19.4. Two time series for the undamped Lotka-Volterra system (19.1) with $r = 1$, $b = -0.5$, $g = -1$, $b = 0.5$ and where the prey exhibits intrinsic exponential growth. The time series with lower peaks corresponds to initial conditions $N(0) = 0.5$ and $P(0) = 2$ and the one with higher peaks corresponds to initial conditions $N(0) = 0.4$ and $P(0) = 2.1$.

population become larger than it would otherwise have. But in one case the effect returns in every cycle, whereas in the other case it dies out.

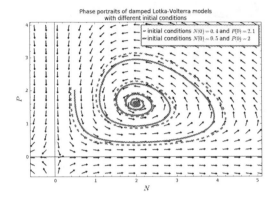

FIGURE 19.5. Two trajectories for the undamped Lotka-Voletta system (19.1) with $r = 1$, $b = -0.5$, $g = -1$, $b = 0.5$, and without the N/K term. The inner trajectory corresponds to initial conditions $N(0) = 0.5$ and $P(0) = 2$ and the outer trajectory corresponds to initial conditions $N(0) = 0.4$ and $P(0) = 2.1$.

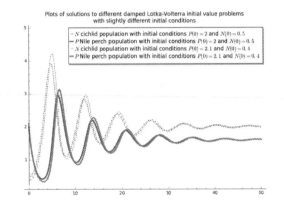

FIGURE 19.6. Two time series for the undamped Lotka-Volterra system (19.1) with $r = 1$, $b = -0.5$, $g = -1$, $b = 0.5$ and where the prey exhibits exponential growth. The time series with lower peaks corresponds to initial conditions $N(0) = 0.5$ and $P(0) = 2$ and the one with higher peaks corresponds to initial conditions $N(0) = 0.4$ and $P(0) = 2.1$.

A population will have different effects depending on where in the population cycle it is done. Sometimes the point of harvesting is to maintain an equilibrium (for example moose license policy in New Hampshire), sometimes it is to maximize yield over time (in the fishing industry for example), and sometimes the point of harvesting is extinction of the species in an area, at least for a while (such as Japanese beetle in my garden or Anopheles mosquito in malaria infested regions). If the predator/prey models in this chapter are good qualitative descriptions of the way the organism grows then they can be used to design optimal strategies for harvesting that organism, even in the face of random disturbances of the system.

19.4. Problems

All of the questions at the end of Chapter 18 remain pertinent here. It is worth revisiting them using a damped version of the predator prey model to see what qualitative or quantitative effects are changed by the assumption of limited growth of prey.

CHAPTER 20

Predator Satiation

What does it mean to eat well? By now we should know what nature has always known. Eating well means eating as much as you need, not as much as you can. For humans this is an important distinction, but for most species the two amounts are often the same. It is efficient to have a stomach capacity that accurately reflects the calories needed for survival and reproduction. A stomach that holds a lot more might reduce the organism's capacity to do other things. Most creatures stop eating when full.

In the last chapter we looked at the growth term for prey in the predator/prey equations and replaced it with a logistic term that limits the growth of prey in the absence of predator. We ended up with these equations:

$$(20.1) \qquad\qquad N' = rN(1 - N) - bPN$$
$$P' = -gP + cNP.$$

In this chapter, we look at the terms that respond to the rate of predation, bPN and cNP. For a fixed amount of prey, N, this term is just proportional to the number of predators, P. This seems reasonable because doubling the number of predators would probably double the chances that a prey organism would be caught.

20.1. Biological context

For a fixed amount of predators, P, the same term is just proportional to the number of prey, N. It says that doubling the amount of prey will result in twice as much prey being eaten. If there are a lot of predators relative to the amount of prey, this would certainly make sense. The predator will eat as much as it can catch. But if there are only a few predators and a huge supply of prey, it makes less sense. The prey would eventually be constrained by its own lack of hunger or the amount of time in the day or some other factor and would be unable to continually double the amount it eats indefinitely. The predator would eat at most as much as it needs. So at low prey levels the rate of predation might rise proportionally but at higher prey levels it would become constant.

This general phenomenon is known as "functional response" and it comes up in a lot of other contexts. For example when your immune system responds to an infection it produces white blood cells in response to the quantity of infection present, but there is a maximum rate at which it can do so. A single heart cell admits calcium ions through molecular gates. At small concentrations of calcium it does so in proportion to the concentration, but at higher concentrations the rate of entry is limited by the number of gates through which the ion may pass. A similar functional response holds for calcium transport.

To take the predator/prey situation as an example, we might think of (20.1) in the slight more general form:

(20.2)
$$N' = rN(1 - N) - bP \times f(N)$$
$$P' = -gP + cP \times f(N),$$

where $f(N)$ is some function of the prey population that starts out looking like "N" but after a while approaches a constant, say "1". So at low populations of prey the term $cP \times f(N)$ is approximately cPN but at high prey populations it is approximately cP. So we need a function $f(N)$ that looks like N when N is small but like 1 when N is large. Put in the language of mathematics, we would say that the limit of $f(N)$ as N goes to infinity is 1, and also the limit of $f(N)/N$ as N goes to zero is 1. Another way of saying the second limit is that "$f(N)$ is on the order of N as N goes to zero" or even "$f(N) = O(N)$ at zero". This means that not only does $f(N)$ approach zero near $N = 0$ but it does so linearly, like N. In Figure 20.1, we can see the graph of such a function.

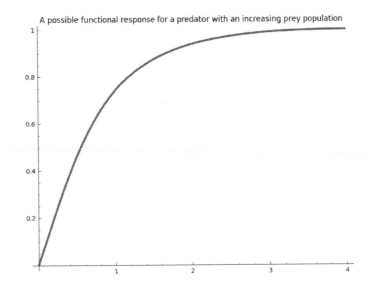

FIGURE 20.1. A desirable functional response for a predator: we observe that for a small number of prey the predation is proportional to the number of prey (in this case the constant is 1) and for a large number of prey it levels off.

There are a lot of functions that can look like this. The simplest such function is $f(N) = N/(1 + N)$, whose graph can be seen in Figure 20.2.

Sometimes our assumptions of predator behavior are more nuanced. Perhaps we believe that if N is very low the predator won't bother to hunt it. This would hold especially if there are alternative prey sources, as in the case of the Nile perch. Then we want a function whose slope is zero near zero but rises to a limit of 1; the graph of such a function can be seen in Figure 20.3.

Functions of the form $f(N) = N^k/(1 + N^k)$ all do this job well. The exponent, which must be greater than 1, controls how quickly the function rises to 1 (see Figure 20.4). When the predator is eating as much as it possibly can, the constant "b"

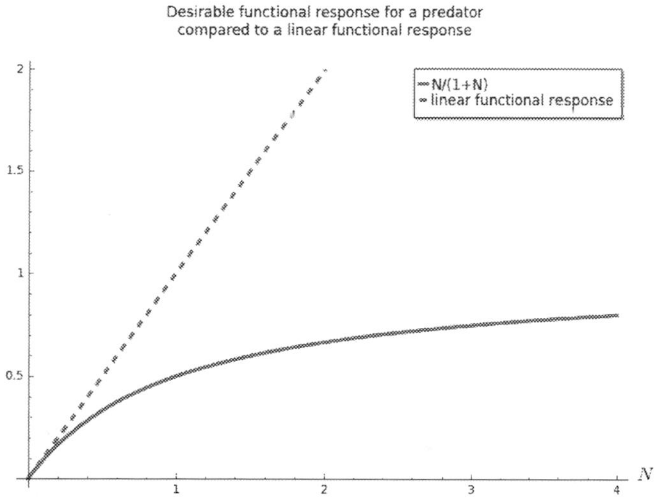

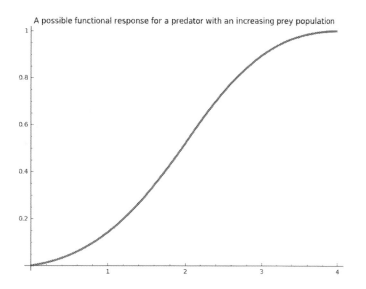

FIGURE 20.2. The graph of $f(N) = N/(1+N)$, a good candidate for a function that starts off as linear but has a horizontal asymptote as $N \to \infty$, modeling the functional response of a predator to a growing population of prey.

FIGURE 20.3. Another possible function response for a predator with alternative prey. In this scenario the rate of predation starts low (maybe the prey are hard to find) but then eventually picks up and then levels off again.

becomes the amount of prey per predator per time unit that perch with unlimited access to prey are able to eat. For Nile perch, one study observes perch to eat once a day, with a full stomach containing 1 to 8 percent of the weight of the perch in prey fish. A modeler would use this study to figure out the constant "b". It can be tricky to estimate this in terms of absolute units such as carrying capacity. The food web supports fewer top predators (in terms of biomass) than prey animals, so one might invoke rules of thumb to get such estimates. The constant "c" then represents the proportion of weight of prey consumed that is actually converted into predator biomass in the next generation.

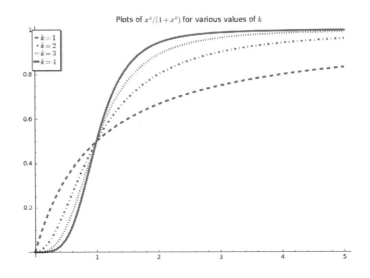

FIGURE 20.4. Graphs of $f(N) = N^k/(1 + N^k)$, for $k = 1, 2, 3, 4$. The larger the exponent, the fast the graph levels off at 1.

Modelers use a variety of functions to capture these different assumptions about predator behavior or other rate-limited processes. In some cases, such as calcium transport across a cell membrane, the process can be measured directly to give both a functional form and accurate constants. In other cases, such as ecology, the form of the function just embodies assumptions about the behavior of an organism and it may not be possible to get constants directly from data.

In the literature these functions are sometimes called "Holling functions" or "Michaelis-Menton functions", depending on the field but usually in ecology or medical literature. Mathematicians do not usually give special names to particular simple functions. They would call f a "rational function" when it is the quotient of two polynomials (when k is a positive integer).

20.2. The model

Ecology models that incorporate a functional response of this sort for predators are sometimes use to model a phenomenon known as "predator satiation." This situation occurs when the prey organism procreates so abundantly that its predators cannot possibly consume a proportional number of offspring. It is usually cited as an adaptive mechanism to avoid predation. To see the adaptive value of this strategy we could compare two models.

- **Model 1**: logistic growth of prey and strictly proportional predation

$$N' = rN(1 - N) - bPN$$

$$P' = -gP + cNP.$$

The time series for the solutions can be seen on the left-hand side of Figure 20.5.

- **Model 2**: logistic growth of prey and predation following a Holling functional response:

$$N' = rN(1 - N) - bP(N/(1 + N))$$
$$P' = -gP + cP(N/(1 + N))$$

The time series for the solutions can be seen on the right-hand side of Figure 20.5.

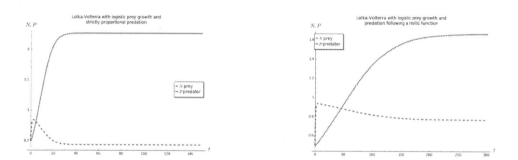

FIGURE 20.5. The time series of Model 1 (on the left) and Model 2 (on the right). The first model is a typical Lotka Volterra model with logistic growth and proportional predation and the second is a Lotka Volterra model but with predation dictated by the Hollings function $f(N) = N/(1 + N)$.

In this case the models are compared in order to justify the statement that predator satiation does indeed have adaptive value to the prey organism. A complete study would vary the growth rate of the prey (the constant r) to see how it affects long term behavior. In Figure 20.6 results for various choices of r for Model 1 are displayed and analogous graphs can be seen in Figure 20.7 for Model 2. We observe that the behavior in Model 2 varies a lot, suggesting that predator satiation does influence the population of prey.

A model of this sort might be used to justify a statement about evolution itself, inferring very long term consequences by the limiting behavior of a system which, although a "long term" consequence of the system, is still a short time frame compared to that of evolution. In Figure 20.5, notice, for example, that switching from model 1 to model 2 increases the equilibrium value for the prey and decreases the equilibrium value of the predator. The constant r does not affect the equilibrium value of prey in either model (see Figures 20.6 and 20.7). The comparison of what happens to equilibrium values of predator is displayed for both models but not very elegantly. A better visual display would be a plot of r versus equilibrium values for each of the two models. The effect of the other constants on these models might also be interesting. Finally, a good modeler would solve explicitly for equilibrium values in both models if possible, which would give an analytic expression for the dependence of equilibrium states on all parameters.

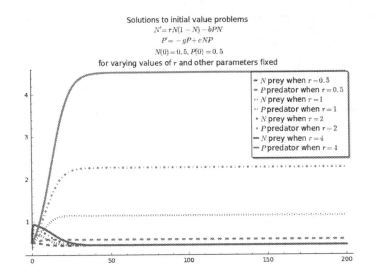

FIGURE 20.6. Model 1 with varying choice of r. The two dashed lines correspond to the predator and prey for one choice of r, the two solid lines as well, etc. The prey all quickly converge to the same equilibria (independent of r) whereas the predators converge to different equilibria, depending on the growth rate of the prey.

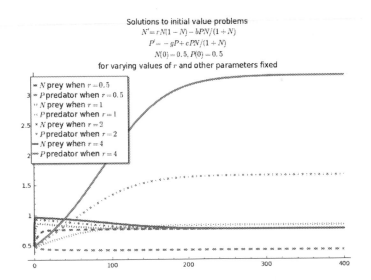

FIGURE 20.7. Model 2 with varying choice of r. The two dashed lines correspond to the predator and prey for one choice of r, the two solid lines as well, etc. The prey all quickly converge to the same equilibria (independent of r) whereas the predators converge to different equilibria, depending on the growth rate of the prey.

20.3. Problems

PROBLEM 20.1. Use data about feeding and reproductive habits for perch and general knowledge about energy transfer in the food web to estimate the constants in the models in this chapter.

PROBLEM 20.2. Some species reproduce seasonally. In this case "r" is not really constant, but oscillating. One could model "r" as an oscillating function of time. Do this.

PROBLEM 20.3. What is a better strategy for prey? A continued high growth rate or an oscillating growth rate? What output of your model is the best measure of "better" in this case?

PROBLEM 20.4. The Nile perch produces offspring as eggs which mature into juveniles and later into breeding adults. The typical life span is around 8 years, with 3 years spent in the juvenile non-reproductive stage. Juveniles eat prawns, mollusks and insects whereas the adults eat cichlids and other small fish. The adults eat the juveniles if they can find them. (According to some sources up to 60 percent of stomach contents of an adult perch have been found to be juveniles. Perch can eat 1-8 percent of their own weight per day.) Maturation of juveniles depends on the number of juveniles and the carrying capacity of the larger system for juvenile perch. It is reasonable to assume the predation of adults on juveniles obeys a Holling functional response. Build a model for this situation where the only prey available to the adult perch is its own juveniles.

PROBLEM 20.5. For the model in Problem 20.2, what parameters give long term continuation of all populations? That is, what (if any) growth and maturation rates will allow the Nile perch to survive while living entirely off its own young?

CHAPTER 21

Isoclines

Africa is the birthplace of humankind, to the best of our knowledge. The original Garden of Eden, she is the land from which all human migration began. With our current scientific understanding, we believe her to be the source of the human race.

Let us turn this observation on its head. From the point of view of the myriad other organisms inhabiting the African continent, Africa is the place humans have inhabited the longest. Her ecosystems are those first to include humans. Her people and her ecosystems have co-evolved for longer than any others on this planet. From Africa came the original human diaspora.

21.1. Biological context

In the area where a new species evolves, the ecosystem evolves along with it. The place where the Nile Perch evolved likely included a complex system of predators, parasites and diseases that co-evolved with it. But when the perch was introduced to foreign waters, such as Lake Victoria, it was in a special position. It may have brought its own diseases and parasites with it, but it left its predators behind. The niche it occupies in a new location used to be taken up by one or more indigeous species that had no competitors until the perch arrived. In particular, the Bagrus Catfish found itself in competition with the Nile Perch for smaller fish.

We would expect a similar but more dramatic scenario when humans arrive in a new location. Some have argued that, in their place of origin, early humans did indeed have natural predators, although these arguments may be rather speculative at this point. In any case, when humans moved out of that first ecosystem and into a new one, their predators probably did not move with them. This situation gave them a competitive advantage with other species. Modern human beings are omnivorous. Although we don't always know what their early eating habits might have been, it is likely that they were strong competitors with indigenous species for a variety of food sources. Some have pointed out that periods of mass extinction in certain locations loosely coincide with the arrival of human beings at those locations. Although this observation proves nothing, it is consistent with our understanding of what happens when species are introduced to a new ecosystem.

Our understanding of this phenomenon rests squarely on mathematical models, as controlled experiments on the scale of whole ecosystems are impossible. Why might the introduction of a new species cause the extinction of at least some of its competitors? Does it always work that way? If not, what circumstances must be in place for both species to co-exist? Our model is built from a simple premise. In isolation, without natural predators, each of the two competing species will grow in population according to the logistic equation we looked at earlier. Let us use P

for perch and C for catfish. Separately, these equations look like this

$$\frac{dP}{dt} = aP(1-P)$$
$$\frac{dC}{dt} = bC(1-C).$$

When we put those two species in the same lake, each of their equations gets a new term added to reflect the disadvantage it has in the presence of the other population. If the perch population is constant, an increase in catfish puts a negative strain on it. If the catfish population is constant, then an increase in perch also puts an extra strain on the perch population, in proportion to the number of catfish present. So, the term we add to both of these equations will be a product, PC, as in the Lotka Volterra model. You might say that P represents the chance a perch will be hunting in a certain area and C represents the chance a catfish is hunting in a certain area. If their behaviors are independent of each other, then PC is the chance both of them are hunting in the same region. That is the situation that results in competition and a disadvantage for both species. The resulting competition equations look like this:

$$\frac{dP}{dt} = aP(1-P) - mPC$$
$$\frac{dC}{dt} = bC(1-C) - nPC.$$

Of course, the disadvantage may not be fair, so in both cases there will be a constant in front of this term, and the two constants may not be the same. Competition between human beings and other top predators provides an excellent example of a situation where the disadvantage is likely to be unequal. When humans enter a new region as hunters they might compete with lynx, let us say, in the taking of smaller animals for food. Now, the lynx is completely dependent on a supply of these small animals for its diet, whereas humans are omnivores. If rabbits are in short supply, the people may rely more on fish. In extremely dire times they may rely completely on vegetables for a fairly long duration. In modern examples, humans as farmers may maintain a supply of their own protected species that the lynx cannot touch. In any case, an increase in humans will have a far larger impact on the lynx than an increase in lynx will have on humans. For the equations describing a competitive interaction, the constants will be very different for the two populations.

As in the Lotka Volterra model, we will analyze these equations using a two-dimensional phase portrait. For these equations, something very different happens. You should remember that, in the Lotka Volterra model, when each of the derivatives is set to zero the necessary consequence was that either one of the populations was zero or else both populations were at fixed point (also called an equilibrium). Now we are going to look more closely at the situation where only one of the derivatives is set to zero. The locations where dP/dt is zero in the equation above look like this:

$$C = \frac{-a}{m}P + \frac{a}{m}.$$

In addition to the line where P is extinct, there is another line in the phase portrait, called an isocline, pictured in Figure 21.1. Because we are plotting P along the horizontal axis, to the right of this line P is decreasing, so trajectories must move to the left. To the left of the isocline P is increasing, resulting in trajectories that move to the right. On the line itself, P is (temporarily) constant.

EXERCISE 21.1. Reinterpret the previous paragraph in terms of the sign of dP/dt. Make a sketch like Figure 21.1 and draw an arrow in each region to indicate in which direction the left-right motion of a trajectory would go.

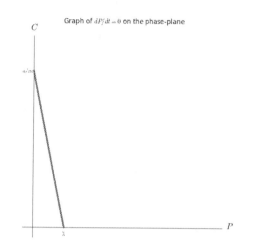

FIGURE 21.1. The graph of $dP/dt = 0$ on the phase plane. In particular, on the slanted line, we see that $dP/dt = 0$.

If we set dC/dt to zero we get

$$C = 1 - \tfrac{n}{b}P$$

which gives a second isocline, shown in Figure 21.2.

EXERCISE 21.2. Reinterpret the previous paragraph in terms of the sign of dC/dt. Make a sketch like Figure 21.2 and draw an arrow in each region to indicate in which direction the up-down motion of a trajectory would go.

These two isoclines may or may not intersect. If they do intersect, we have an equilibrium point, shown in Figure 21.3. If not, we see something like Figure 21.4.

EXERCISE 21.3. Explain in terms of dC/dt and dP/dt why the intersection of the two isoclines corresponds to an equilibrium point.

Because of the different roles of the two lines in determining direction of motion, we actually get four scenarios, depending on the constants in our equations, shown in Figure 21.5.

For each of these possibilities we only need to think about trajectories that begin inside the rectangle determined by the two limiting populations, because initial populations above the carrying capacity are unlikely. So, for the first scenario in Figure 21.5, we are looking at a box containing three regions whose long term behavior may be different. We need to look at each of these separately. They are labelled A, B, and C in Figure 21.6.

In the region labelled A, all trajectories are moving downwards and to the left. It is easy to see that they will eventually have to enter region B. In region C, all

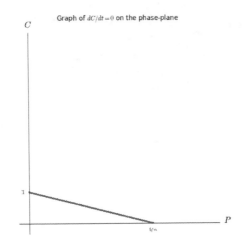

FIGURE 21.2. The graph of $dC/dt = 0$ on the phase plane. In particular, on the slanted line, we see that $dC/dt = 0$.

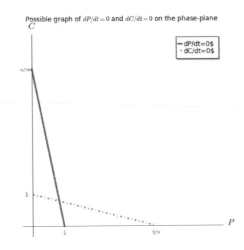

FIGURE 21.3. An equilibrium point where isoclines cross.

trajectories are moving upward and to the right. It is easy to see they will also enter region B eventually. In region B, however, the trajectories are moving down and to the right. They cannot enter region C because when they get to the isocline all downward motion stops and they are forced to the right, back into B. For a similar reason, they can't enter A either.

Within B, they move forever downward and to the right. The catfish are dying out and the perch are approaching their limiting population. All roads lead to the fixed point at the intersection of the horizontal axis and the higher isocline. In this scenario, the perch will outcompete the catfish and drive it to extinction, no matter what the relative populations are when the perch is introduced.

This kind of analysis can be done for each of the four scenarios.

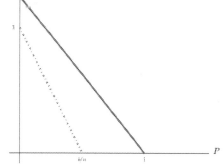

FIGURE 21.4. No equilibrium point inside the quadrant.

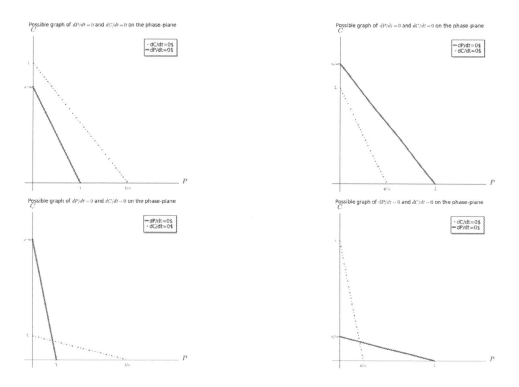

FIGURE 21.5. All four possible options for the placement of two lines in the first quadrant both with positive slope.

EXERCISE 21.4. Carry out the analyses in the other three cases. Identify which species survive and the equilibria the system tends to. Check that it is consistent with the following paragraph.

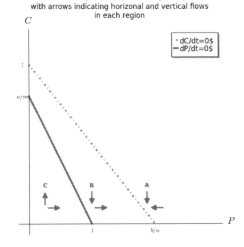

Isoclines of dP/dt and dC/dt on the phase-plane
with arrows indicating horizonal and vertical flows
in each region

FIGURE 21.6. Isoclines for the first case. The arrows correspond to when $dP/dt > 0$ (the horizontal arrows pointing to the right), when $dP/dt < 0$ (the horizontal arrows pointing to the left), when $dC/dt > 0$ (the vertical arrows pointing up) and to when $dC/dt < 0$ (the vertical arrows pointing down).

The reader will find, after having completed the exercise, that in the first two of the scenarios one species is always forced to extinction. In the third scenario both species co-exist and come to equilibrium. In the last scenario, the final outcome depends on the equilibria of the original populations of both species.

Our example illustrates several important ideas. First of all, we see the importance of thinking through simple algebraic consequences of our model, as they apply to the phase portrait for it.

Second, when analyzing a complex system using time series output from the computer, one must try a variety of initial conditions. If the phase portrait lurking behind the numerical computation has initial conditions like the fourth above, there will be very different conclusions depending on what the populations were at the beginning of the calculation.

Third, we see how important it is to have an idea of what the constants are in our equations. A slight change in the constant might have enormous long term effects on the system, as illustrated in Figure 21.7.

Fourth, we see that the mathematics gives us enormous insight into natural systems. To the extent that we believe our equations are a fair representation of the relationships between organisms, the phase portrait analysis vividly portrays the range of possible outcomes and the factors upon which each outcome depends. This is useful information for the purpose of both predicting outcomes in an ecosystem and also for interfering with those outcomes. Any modification in behavior of one of the two organisms in our example will affect one or more of the constants involved. Human behavior could also be added to this example, such as fishing patterns that favor catching one species over the other. In fact, human fishing patterns on the lake are tailored to the catching of perch, which is highly exportable, rather than catfish.

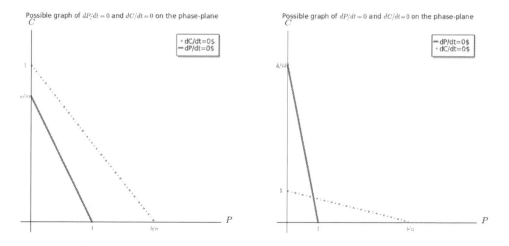

FIGURE 21.7. A change in a/m: we get very different behavior if $\frac{a}{m} > 1$ or $\frac{a}{m} < 1$.

The reader is invited to investigate how such behavior would alter the constants in the above equation so that the Bagrus catfish, although low in population, may be prevented from extinction. These models, and their interpretation, are the only predictive guides we have to intervention in the large, complex systems that routinely arise in ecology, physiology, and epidemiology.

21.2. Problems

PROBLEM 21.5. In which of the four possible scenarios does a species always become extinct? In which case do both species always survive? Draw a picture of a trajectory in this case. What are the constraints on a, b, n, m? How would you explain the physical meaning of the parameters in these equations? What might affect them?

PROBLEM 21.6. One case has different outcomes depending on the initial position. Which is it? Draw the phase plane and indication which regions lead to which long term behaviors of the population. What are the implications of this in terms of ecological intervention? The general term for this sort of phenomenon is "population dependent effects", because the long term behavior of the system is dependent on the starting populations (unlike the other three scenarios).

PROBLEM 21.7. There are two scenarios that lead to extinction of one species or the other. As the parameters change, one morphs into the other. Is it possible to perturb the parameters in such a way that there is no time at which coexistence is possible? Explain the implications of your answer in terms of ecological intervention.

CHAPTER 22

Species Formation

Five hundred distinct species of cichlid fish are endemic to Lake Victoria. The word "endemic" means that they are found there and nowhere else. In the case of Lake Victoria cichlids, this is not strictly true as they are a popular freshwater aquarium fish in tanks around the world. But their only natural habitat is Lake Victoria. Other nearby tropical lakes boast a similar diversity of (different) cichlid species. Geologists believe that Lake Victoria suffered a drought about 14,700 years ago, drying out completely. As it refilled, some ancestors of the current cichlids must have arrived. If the current diversity of the lake were derived from a single originating species, then the rate of evolution of individual cichlid species would be about one new species every 30 years.

Biologists dispute the 14,700 year geological estimate, based on DNA evidence in the cichlid population. They theorize that the lake did not dry up completely, preserving the cichlids in a series of disconnected ponds. If we disregard the drought of 14,700 years ago, the previous period of complete dryness would be about 100,000 years ago. This estimate would result in a new species appearing approximately every 200 years. Furthermore, the hypothesis of preservation of the species in many disconnected ponds over an extended period provides a mechanism for evolution: the process of isolation and reintroduction of species.

When a population of species is cut into two parts that can no longer interbreed, the genetic makeup of the two separate cohorts will slowly drift apart. Mutations are random, and a sequence of mutations in the DNA of a population causes the genetic makeup of that group to vary from generation to generation. A mathematician would say that the mutations are causing the genetic composition to take a "random walk", with slight changes accumulating to cause, in the end, a real divergence of genetic makeup between the two cohorts. Of course, this process occurs in response to changes in the rest of the ecosystem. In the case of fish in two newly separated ponds, all the other species in the pond undergo the same process. The population of all organisms in the pond is a linked ecosystem coevolving.

Ultimately enough difference accumulates between the two fish populations that they can no longer interbreed. It may take only a small difference: a missing visual cue, a slight change in breeding cycle. If the two populations are then mixed, the two populations may refuse to hybridize at all, remaining separate breeding pools. Sharing the same habitat, they now compete for resources. We can use the model from Chapter 21 to investigate the claim that isolation followed by reintroduction is a viable way of developing new species.

As we saw in Chapter 21, it could easily be the case that one species is driven to extinction. In fact, in 3 of the 4 cases we looked at, this was indeed the case. In other words, if the parameters of the equations for the competition model were chosen at random, extinction of one of the two populations would be the most likely scenario. Of course, it is not the case that the parameters are random in the case of isolation and reintroduction. Rather, a few random mutations have occurred that may or may not alter each of the parameters in the competition equation.

Many mutations are inconsequential to survival. Most of the features we use to distinguish ourselves from each other as humans are not really important features from an evolutionary perspective. So perhaps the first question we should ask about our two species of fish is what the competition equations will look like if the two species are identical in every way that might affect the four parameters in those equations. In this case we get two identical equations:

$$X' = aX(1 - X - Y) = aX(1 - X) - aXY$$
$$Y' = aY(1 - Y - X) = aY(1 - Y) - aXY.$$

The assumption here is that both X and Y were growing according to the logistic equation in separate compartments until reintroduced. At this point they share resources. We have choice as to whether we want to consider the resource available to the combined population as still equal to 1 or whether we assume it is now doubled. In any case, if we assume both populations start at the same size, we see the result in Figure 22.1.

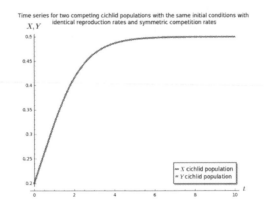

FIGURE 22.1. The time series for the two cichlid populations starting with the same initial condition ($X_0 = Y_0 = 0.2$) and with $a = 1$. One time series lies on top of the other.

On the other hand, it is possible that one of the isolated habitats was smaller than the other, in which case one population (say X) starts out with fewer members than the other. In that case, what happens? See Figure 22.2 for an example of what would happen in this case.

As we can see, when neither nature nor mathematics can distinguish between the populations, both persist. The difference in initial values, however, determines the long range proportions of each in the combined system.

Now that we know what happens if there is no significant mutation, let us see what happens if the populations diversify enough to affect either growth rates or the effect of competition. First let us suppose that a mutation improves the growth rate of one of the two populations, without affecting the way they share the

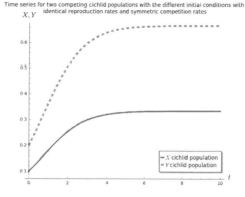

FIGURE 22.2. Different starting values for X and Y (in particular $X_0 < Y_0$). We continue to let $a = 1$.

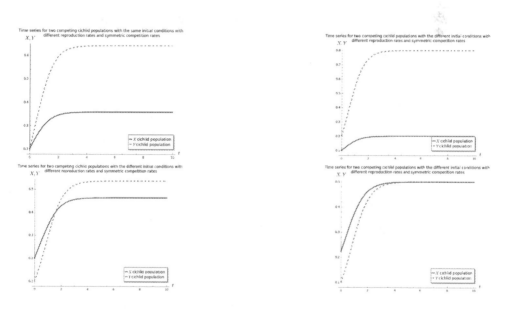

FIGURE 22.3. Time series with $a = 1$ and $b = 2$ but four different sets of initial conditions. We observe that coexistence is possible.

resources. This would yield two equations of the form:

$$X' = aX(1 - X - Y) = aX(1 - X) - aXY$$
$$Y' = bY(1 - Y - X) = bY(1 - Y) - bXY.$$

Now what happens? We can see from the following runs with $a = 1$, $b = 2$, and various starting values of X and Y in Figure 22.3.

As we see, coexistence is still likely. Now we can ask what sort of effect a mutation might have on the way the two species interact. If the two species develop slightly different food preferences, this effectively reduces the competition between them. Keeping the basic growth rate the same, we then have two equations that will look like this:

$$X' = aX(1 - X) - mXY$$
$$Y' = aY(1 - Y) - nXY.$$

Comparing this with the system for identical species, our hypothesis would require that m and n both be slightly less than a.

EXERCISE 22.1. Why is it the case that both m and n should be slightly less than a?

In other words, species Y has a less detrimental effect on the habitat of X than does X itself, and vice-versa. The analysis we did on phase portraits in the last chapter tells us which situation we see in this case. The populations tend to equilibrium values. In this case the equilibrium is an "attracting state", which just means that populations starting in the neighborhood of the equilibrium all end up there. Notice this is unlike the situation above with identical populations, where the equilibrium value depending on the initial populations of each type. In Figure 22.4 you can see an example of this type of long term behavior.

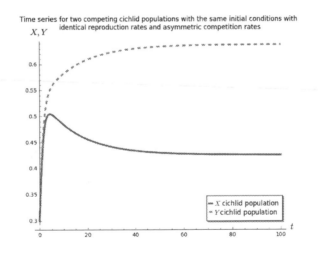

FIGURE 22.4. $X_0 = Y_0 = 0.3$, $a = 1$, $m = 0.9$, $n = 0.85$.

Ecologists have observed that even closely related species such as the cichlids of Lake Victoria and Darwin's finches often have slightly different diets, claiming that diversification of niche allows the species to cohabit successfully. Even zooplanktons have been observed to differentiate themselves via food preference. The simple competition models support the hypothesis that niche differentiation enhances the ability of species to coexist. In these models even a slight decrease in the impact each population has on the other one is enough to guarantee the presence of a stable, attracting equilibrium with populations of both species present. The models are also consistent with the hypothesis that isolation followed by reintroduction of species into contact with each other is a cause of increasing diversity of related species.

22.1. Problems

PROBLEM 22.2. Of course other mutations are also possible, including ones in which one population becomes a slightly stronger competitor, perhaps catching prey more efficiently than the other. In this case either m or n may be slightly larger than the constant 'a'. What happens in this case?

CHAPTER 23

Top Predators

The presence of top predators can have surprising impact on the survival of species it preys upon. In this chapter we investigate some of these impacts.

23.1. Biological context

In Chapter 22 we looked at the process of isolation and reintroduction of populations to see how the competition model explains much of why that process works to produce new species. However there are in nature examples of species that not only compete with one another but are also both prey for some predator in a higher trophic level. Lake Victoria's cichlids are certainly one example of this, as all the many species of cichlid are prey for both Nile perch and Bagrus catfish. Although the perch would not yet have been introduced at the point where previously isolated populations were reintroduced, the catfish would have been present. It is possible to combine the models for predator-prey and competition to look at models that describe this structure: see Figure 23.1 for a visual representation of this model.

Looking at the simplest example with one predator and two prey, we would expect three differential equations to govern this system.

23.2. The model

What equations should govern this system? For some situations, the two-prey species share a food source or habitat, so we might expect that, in the absence of predator, the quantities X and Y would satisfy some kind of competition model. If we think this is the case we would add a relation between X and Y to the box model above, and express that relation as a competition:

$$X' = aX(1 - X) - mXY$$
$$Y' = bY(1 - Y) - nXY.$$

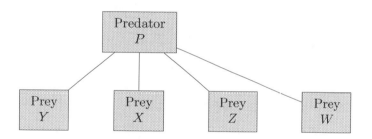

FIGURE 23.1. A box model for a predator with four prey species that do not interact among themselves.

189

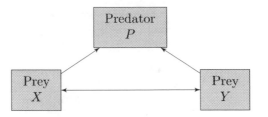

FIGURE 23.2. A box model for a predator with two species that compete with each other.

In the absence of Y, we would expect X and P to satisfy a predator prey relation:

$$P' = -kP + cPX$$
$$X' = aX(1 - X) - rXP.$$

And similarly, in the absence of X, P and Y should satisfy a predator prey relation:

$$P' = -kP + dPY$$
$$Y' = bY(1 - Y) - sYP.$$

Looking at these three requirements tells us that the three organisms together should satisfy:

(23.1) $$P' = -kP + cPX + dPY$$
(23.2) $$X' = aX(1 - X) - mXY - rXP$$
$$Y' = bY(1 - Y) - nXY - sYP.$$

It is worth pausing here to notice that this analysis of our system, consisting of articulating hypotheses about pairs of organisms and then stating these mathematically, is the right way to approach complex systems. It is the only way to guarantee that the simpler components of the system behave correctly when the other parts are absent. We have to guarantee this in case one of the organisms goes to extinction in our complex model, which we will see is sometimes the case.

In Figure 23.3 we see several possible outcomes for this system, with various parameter choices. Although each predator-prey pair has a coexistent equilibrium when the second prey species is missing, when all three are put together one of the two prey species may be driven to extinction.

23.3. Analysis of the model

If we want to know what possible equilibria look like, we can solve for them using these three equations:

$$0 = -kP + cPX + dPY = P(-k + cX + dY)$$
$$0 = aX(1 - X) - mXY - rXP = X(a(1 - X) - mY - rP)$$
$$0 = bY(1 - Y) - nXY - sYP = Y(b(1 - Y) - nX - sP).$$

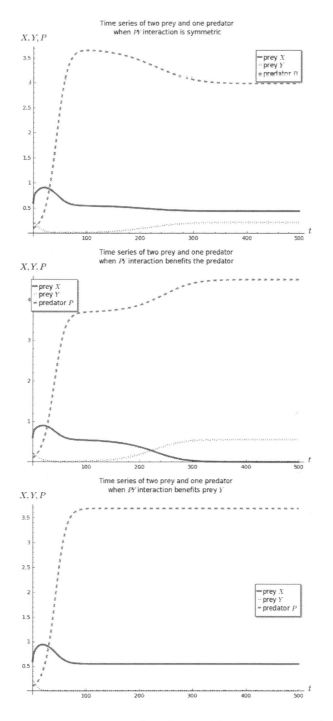

FIGURE 23.3. Time series for the model represented in (23.1). The parameters $k = -0.11$, $c = 0.2$, $a = 1.0$, $m = 0.9$, $r = 0.122$, $b = 1.0$ and $n = 1.1$ are the same in all three graphs. The first graph has $d = 0.1$ and $s = 0.1$. The second has $d = 0.2$ and $s = 0.1$. The third has $d = 0.1$ and $s = 0.5$.

We certainly get equilibrium at extinction $(X = Y = P = 0)$ and we also get other equilibria when we set one or two of the quantities equal to zero. Here are all possible equilibria triples where one or more quantities is zero:

- $X = Y = P = 0$
- $X = P = 0, Y = 1$
- $Y = P = 0, X = 1$
- $X = 0, Y = k/d, P = b/s(1 - k/d)$
- $Y = 0, X = k/c, P = a/r(1 - k/c)$
- $P = 0, X = (a/m - 1)/(a/m - n/b), Y = (b/n - 1)/(b/n - m/a)$, valid where positive

EXERCISE 23.1. Check these equilibria and explain why there cannot be any more equilibria where at least one population is extinct.

In all of these cases, one or more populations is extinct. However there is a possibility of coexistence when X, Y, and P are all nonzero solutions of:

$$0 = (-k + cX + dY)$$
$$0 = (a(1 - X) - mY - rP)$$
$$0 = (b(1 - Y) - nX - sP).$$

EXERCISE 23.2. This system of equations come from the previous system of equations. How did we get from the previous system to this one and why is that allowed?

These three linear equations can be solved by software or brute force. For some parameter choices the solution will yield three positive numbers for X, Y, and P, which are biologically valid solutions. However, for some parameter choices the equilibrium will be unstable and therefore will never appear in simulations. For other choices the equilibrium will indeed be stable.

This situation raises the following question. Could it be that the presence of a top predator is a prerequisite to species diversity in some cases? In other words, is it possible that nature could have species that satisfy these criteria:

(1) When left to themselves, one species always forces the other to extinction
(2) But will coexist in the presence of a suitable joint predator?

We can test this question with our model. Criterion 1 means that, according to the analysis in Chapter 21, either b/n or a/m is less than 1 and the other is greater. Suppose b/n is less than one and a/m is greater than one. Then X survives and Y becomes extinct. Here is an example of such a pair:

$$X' = X(1 - X) - .9XY$$
$$Y' = Y(1 - Y) - 1.1XY.$$

We can think of this situation as two recently divergent species of cichlid, one of which will surely outcompete the other upon reintroduction. Then we can introduce a top predator, P, with preferential feeding habits. The equations below represent such a predator, having a slightly more detrimental effect on the population of X than on Y.

$$P' = -0.11P + 0.2PX + 0.1PY$$
$$X' = X(1 - X) - 0.9XY - 0.122XP$$

$$Y' = Y(1 - Y) - 1.1XY - 0.1YP.$$

We can see in Figure 23.4 the contrast in outcomes depending on whether there is predator present or not.

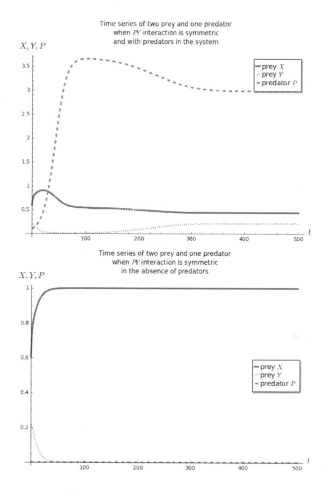

FIGURE 23.4. The top time series is the time series of the populations modeled via the equations in this section when $P_0 \neq 0$. We note that no species go extinct. On the other hand, if $P_0 = 0$ we get extinction, as seen in the bottom time series.

These graphs vividly illustrate the beneficial effect of such a predator on the survival of a species. If an ecosystem is behaving according to this model, removal of the predator will cause one of the prey species to go extinct.

When drawing conclusions from a model of this sort, we should worry about whether the example given above to illustrate a particular principle is just a very special case. In other words, if we vary the parameters in these equations a bit (but still satisfying the criteria we wanted), will the phenomenon disappear? This question is one of "robustness", which refers to the ability of a system to keep the same qualitative outcome under small perturbations of parameters. There are

9 parameters in the above system of equations, which represent the 9 constants that were chosen to illustrate this phenomenon. Two key constants represent the preference of the predator for consuming X over Y. These are the coefficient of PX in the first equation and the coefficient of PX in the second equation.

The coefficient of PX in the second equation can be varied, and we can see in Figure 23.5 the response of the system as it increases from 0.2.

Note that, although the different in equilibrium values for the two-prey species increases, they both persist. Yet in all of these cases, if the predator is removed, Y goes extinct. The contrast can be quite dramatic, as in this case where Y goes from being the predominant prey to extinction as in Figure 23.6.

However if the coefficient of PX in the second equation drops to 0.1 so that the predator has no more detrimental effect on X than on Y, Y goes extinct, as in Figure 23.7.

We would expect Y to go extinct in this case because we have set up equations where X naturally drives Y to extinction. But the large range in which X and Y coexist as we change this parameter tells us that the system is robust with respect to that parameter. For a full study of robustness, we would test all of the parameters in this way. Robustness of the system gives us confidence that a real system would behave this way, because a real system is subject to small random perturbations. Our model would represent these as small perturbations of the parameters in the equations, and we want our qualitative conclusions to remain valid under these sorts of small changes.

One way to think of this collection of models is as support for a hypothesis. In this case the hypothesis is "The existence of a top predator can make coexistence possible for species that otherwise could not coexist". We restated this in mathematical form, "If the hypothesis is true we should be able to find a model where two-prey species coexist in the presence of a predator and do not coexist when the predator is removed". We tested the hypothesis by constructing equations representing two species that would not coexistence in competition, plus a predator that interacted with each of them. We found parameters that satisfied both requirements, and we showed that the system is robust under small changes in these parameters (actually we only demonstrated how to do this with one of them). Now we can be far more confident that our hypothesis is correct, because when we quantify it properly it holds true. This example is an excellent illustration of how models are used to test concepts, even when the real parameters are not known.

23.4. An example from the literature

Ecological concepts are very difficult to test in nature, because we don't care to experiment on our own ecosystem. But sometimes experiments just happen anyway. There is a wonderful account, described by E.O. Wilson in "The Diversity of Life" [8], of the removal of sea otters from the Pacific coast of California, Oregon and Washington in the early part of this century. The sea otters were hunted nearly to extinction for their warm pelts, removing a top predator from the ecosystem of giant kelp beds up and down the coast. These ecosystems are abundant and rich, far too complex to model each species. However they are all based on the abundant algae in one way or another. Furthermore, there is one species, the sea urchin, which is only eaten by sea otters. It has no other predator and is not linked to the other species,

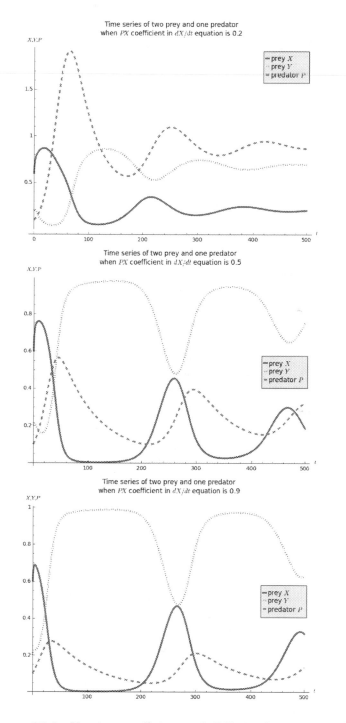

FIGURE 23.5. Varying coefficients of PX in the equation for dX/dt in the model.

FIGURE 23.6. On the left, Y is the dominant prey yet on the right it goes to zero with the predator removed from the system.

FIGURE 23.7. For a sufficiently small coefficient of PX in the $\frac{dX}{dt}$ equation of system (23.1), Y goes to extinction.

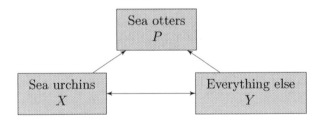

FIGURE 23.8. A box model for the system studied by Wilson [8].

except through competition for algae. So we could lump a large complicated group of interacting species into one biomass (labelled below "everything else"), which is in competition with sea urchins. Both of these lose biomass to the top predator, the sea otter. Our box model looks like the one we have been discussing in this chapter, as seen in Figure 23.8.

The discussion in this chapter would lead us to believe that, with the otter removed, there are three possibilities. It is possible that both prey boxes continue to coexist, or that one is driven to extinction.

What actually happened? Not so long after the otters were removed the diversity of the kelp forests disappeared, and they were described by witnesses as "sea urchin deserts." The lowly urchin managed to outcompete its complex diverse competition. Some years later when the otter was reintroduced, the diversity of the kelp forests returned. Rarely do we see an ecology experiment on such a grand scale. For a convincing explanation of why the simple ecosystem consisting of algae/urchin might be the winner in competition with a complex, highly developed trophic web, read "Why Big Fierce Animals Are Rare" by Paul Colinvaux [3]. The answer he gives is, although free of equations, a deeply mathematical discussion.

The concepts of a top predator making it possible for other species to exist plays out not only in lake Victoria, and other water areas, but also on land. Large animals have a big effect on their environment, irrespective of what the environment is. These concepts can be examined therefore through the relationship between large African Herbivores and their habitat. For general information and conceptual information with data, see [2, 1, 4, 5, 6, 7]. In addition to the competition among herbivores described in these articles, one might also explore the effect of introducing or removing a top predator.

23.5. Problems

PROBLEM 23.3. Is it possible to have three a combination of three competing species, X, Y, Z that always results in coexistence of X and Y and extinction of Z, yet in the presence of a common predator, X and Z survive and Y is driven to extinction? If you can figure out a likely reason why such a thing might happen, you can test it by embodying your reason in a set of four equations. Then you can test to see if it is possible.

PROBLEM 23.4. In regard to Problem 23.3, you might think the competition equation for X looks like this:

$$X' = aX(1 - X) - bXYZ.$$

There is a good reason why this is not the case. Can you figure out what it is?

Bibliography

[1] Alan Birkett. The impact of giraffe, rhino and elephant on the habitat of a black rhino sanctuary in kenya. *African Journal of Ecology*, 40(3):276–282, 2002.

[2] Alan Birkett and Barry Stevens-Wood. Effect of low rainfall and browsing by large herbivores on an enclosed savannah habitat in Kenya. *African Journal of Ecology*, 43(2):123–130, 2005.

[3] Paul A Colinvaux. *Why big fierce animals are rare: an ecologist's perspective.* Princeton University Press, 1979.

[4] Joris PGM Cromsigt, John Hearne, Ignas MA Heitkönig, and Herbert HT Prins. Using models in the management of black rhino populations. *Ecological Modelling*, 149(1):203–211, 2002.

[5] Robin L Mackey, Bruce R Page, Kevin J Duffy, and Rob Slotow. Modelling elephant population growth in small, fenced, South African reserves. *South African Journal of Wildlife Research*, 36(1):33–43, 2006.

[6] AO Nicholls, PC Viljoen, MH Knight, and AS Van Jaarsveld. Evaluating population persistence of censused and unmanaged herbivore populations from the Kruger National Park, South Africa. *Biological Conservation*, 76(1):57–67, 1996.

[7] Johan Van de Koppel and Herbert HT Prins. The importance of herbivore interactions for the dynamics of African savanna woodlands: A hypothesis. *Journal of Tropical Ecology*, pages 565–576, 1998.

[8] Edward O Wilson. *The diversity of life*. WW Norton & Company, 1999.

CHAPTER 24

Modeling Interlude: Potential Problems with Models

Now that you are starting to discover your own approach to modeling biological systems, we think it makes sense to talk some philosophical issues that might arise as you develop more of your own models. These are covered in a variety of general books on modeling, but our discussion is informed mostly by [2] and [1].

24.1. Models can't do everything

In the seminal paper "The strategy of model building in population biology" by Levins [3], it is argued that models have three objectives in mind:

(1) realism,
(2) precision, and
(3) generality.

Moreover, Levins argues further that a model cannot accomplish all three simultaneously. He defines three types of models: Type I models sacrifice generality for the sake of the other two (e.g., the model's utility only depends on knowing all the constants involved), Type II models sacrifice realism for the sake of precision and generality (e.g., many simplifying assumptions are made about the behavior of some of the system's components) and Type III models sacrifice precision for the sake of generality and realism (e.g., one makes qualitative assumptions and deduces qualitative conclusions). In general, this trichotomy is not the final word in a taxonomy of models, but we think it does provide a useful way to conceptualize building models; in particular, it gives license to not have to strive for a perfect model.

Moreover, a widely held misconception is "A more detailed model makes for a better model." One reason that this statement cannot be true is logical; mathematicians tend to think of the extreme cases. Consider: the most detailed model of a system S is itself and while using S to represent itself will give the most accurate result, it will be unwieldly and not very general. So, in turn, the simplification encoded in the decisions made by the modeler, is a virtue of a model. On the one hand, a model should handle the specific purpose for which it was designed and does not need to include every possible aspect of reality. On the other hand, models are designed and implemented by humans whose time is finite. Knowing exactly how complicated to make the model so that it is apt for its purpose is the hardest part of modeling. We therefore recommend an iterative approach: start with a bare-bones models with standard assumptions and add biologically relevant things to the model to make it better represent what has been observed.

24.2. Models can be misapplied

Models are designed to address and develop our understanding of particular problems. There are many ways in which a model we develop can be misapplied, but an important one to keep in mind is that a model is only as good as the data used to develop it. The trouble comes from using a model built upon poor data to make quantitative conclusions. To many people, numerical information seems to be imbued with more truth than verbal information, and so a modeler needs to be careful to make far-reaching conclusions if the model is built on a shaky foundation. In addition to being built on poor data, a model can be based on faulty understanding of the mechanisms at play. If the interactions between parts of a system being modeled are not completely understood, a modeler should again be hesitant to apply the model to make far-reaching conclusions.

24.3. Models can be unverified

Many mathematical models cannot be solved analytically; most of them are solved numerically using methods like Euler's method. Such algorithms are best suited for certain kinds of equations and not all suitable for other kinds. If a modeler is using a set of computer tools or implementations of known algorithms, they should know enough about how the algorithm is implemented and how it works to trust the output the model produces. This is hard in practice because, for example, many modelers use closed-source implementations of algorithms; there is no way for a modeler to check a closed-source implementation only the company that implemented the algorithm has access to the source code. We point out that Sage is open source.

24.4. Models can be poorly calibrated

The estimation of the parameters needs to be done carefully and in a robust way. How can the initial conditions be determined? How about the rate constants? Typically this is done using some statistical techniques (e.g., linear regression) and has to be done well. The data to which the statistical is being applied also have to be collected with a great deal of precision. The data themselves also have to be the right kind of data. For example, if one is measuring the growth of water hyacinth in a lab at a certain temperature and light exposure, how applicable to populations in the field are these lab measurements?

24.5. Models cannot be validated

It is possible that the output of the model cannot be validated or corroborated. There are several reasons this may happen but one is a lack of data. The data required to develop a model of a complex biological system can be very costly (in terms of time, required expertise, cost, etc.) to acquire. Modelers often use all the data at their disposal to make their model. How, then, can the quality of the model be judged? We mean the quantitative quality, in particular. In order to judge this, the same model would have to be applied to another data set. Usually, such a data set is not available to either the modeler or the person trying to validate the model.

24.6. Models are still valuable

We hope that this section does not discourage novice modelers from continuing to develop their skills. The above is meant to illustrate some of the things of which you should be aware as you try to convince your peers of the value added by your model. As long as you understand the limitations of the conclusions that you can reach with your model and appreciate the value of the conclusions that you do reach (limited though they may be), you will be a successful modeler.

Bibliography

[1] David J Barnes and Dominique Chu. *Introduction to modeling for biosciences.* Springer Science & Business Media, 2010.

[2] James W Haefner. *Modeling biological systems: principles and applications.* Springer Science & Business Media, 2012.

[3] Richard Levins. The strategy of model building in population biology. *American scientist*, 54(4):421–431, 1966.

CHAPTER 25

Research Interlude: Making Figures

Figures and tables are important ways to communicate your results to a reader. A poorly designed figure or table can do more harm that a poorly worded sentence. In this chapter, we point out some features of figures and tables as they are used in research papers (the role of figures and tables in a textbook is slightly different and so this text does not adhere to all the rules below).

25.1. Terminology

A figure is a visual representation of results. Examples of figures are graphs, diagrams, charts, maps, etc. Graphs are perhaps the most common figure as they highlight trends of data as well as other relationships that might exist between data. We hope the reader has found the large number of graphs in the text up to this point a useful way to help understand the content of the text.

A table is a way to organize a collection of data. Some data are better suited for figures (e.g., do not use tables to show trends), but tables are useful if you are interested in reporting data from other sources or data that are taken as input into your model but are not the output of your model.

Sometimes, if the data are compact enough, it is preferred to state the conclusion in a simple line of text.

25.2. How to tell if your figure or table is complete

A figure or table must be able to stand on its own. A reader will only read the results section of a paper, say, if they can understand and appreciate what is in the figure or table. The parts of a figure that help with this are:

- its caption,
- its legend, and
- its labels.

Normally it is advised not to include a title in a figure because the caption should completely explain the figure.

About figures. The following are true of a useful caption for figures:

- the first sentence should serve as the title of the figure;
- the source of the data needs to be included;
- the meaning of the trends or relationships in the data in the context of the research question should be explained;
- any appropriate summary statistics should be reported; and
- an explanation of any subtlety in the trends or relationships visible in the figure.

Do not simply restate the axis labels with a "versus" written in between–such a statement would be evident if the labeling of the figure is done correctly.

The labels of the figure are used to indicate what is on each axis, the units of the numbers on each axis and the scale of the data (in the units). The text that describes what is on each axis should not be overly technical (e.g., "time (d)" or "population (1000s)") but needs to include the units. The axes should typically have tick-marks to help the reader approximate points on the graph for which they do not have exact values.

The legend of a figure is useful if there is more than one set of data being represented in the same figure. The legend gives the reader a way to identify which data set corresponds to which part of the figure.

If the labels, legends and captions are used correctly, the figure should be self-contained and not require the reader to have read any of the paper (but, obviously, the reader should understand what the paper is generally about).

About tables. Tables, in general, have fewer parts. The important parts of a table are

- its caption,
- its column titles and
- its footnotes.

Again, useful captions have some standard features:

- the first sentence should serve as the title of the table;
- the source of the data needs to be included;
- an explicit justification of why this table is being included in the paper.

The column titles tell the reader what data are being organized. When appropriate the units of the data should be included in the column title. There should be drawn, solid lines between the caption, the column titles, the data, and the footnotes (not between each line of data, however) to help the reader organize what they are seeing.

Footnotes in a table can serve any one of several purposes. First, they can be used to signify statistical significance–such footnotes might look like 3.3** and down in the footnote area, below the data, it might say "** significant at $p < 0.10$." Second, they can be used to clarify points in the table (e.g., maybe data for the same species is reported twice, a footnote could explain why). Third, footnotes can be used to convey information that is both ancillary to the main point of the table and repetitive (e.g., maybe the table represents populations from two different places–a footnote could be used to indicate which row of the table is from which place).

25.3. Parting advice

We present the above to provide some guidance on what we think makes for a good table or figure. Instead of memorizing these rules, however, we recommend you find a paper that you think does this well. That is, find a paper in which you think the tables and figures (and equations) are wholly self-contained. Identify what you like about them and what gives you the impression that they are self-contained. Keep that paper nearby when you start writing. For the kinds of figures, tables and equations the models in this book require, we recommend papers by Andrew Edwards and John Brindley [1, 2]. Eventually, you will develop your own style of figures and you will no longer have to look to those papers for guidance (in fact maybe you can just look at your previous papers).

Bibliography

[1] Andrew M Edwards and John Brindley. Oscillatory behaviour in a three-component plankton population model. *Dynamics and Stability of Systems*, 11(4):347–370, 1996.

[2] Andrew M Edwards and John Brindley. Zooplankton mortality and the dynamical behaviour of plankton population models. *Bulletin of Mathematical Biology*, 61(2):303–339, 1999.

CHAPTER 26

Projects for Predatory-Prey Models

26.1. Biological context

In the heart of the Brazilian rainforest, nestled in a marshy floodplain far from the intrusions of 19th-century mankind, a beautiful floating plant existed, clustered together with seven generations of its own daughters. The plant, a water hyacinth (*Eichhornia crassipes*), was not especially remarkable, for it was only six inches tall. If one were to plunge under the unruffled surface of the swamp, the anchoring system of the uniquely crafted boat-like herb would be visible. Long, purple, feathery roots trailing several feet or more, anchoring each individual plant to the clayey sediments coating the bottom; runners extending from the mother plant to each of several hundred daughters, which budded without sexual reproduction from those long extensions–all this would be visible, as well as schools of small, colorful fish and squirming insect nymphs seeking shelter amongst the feathery roots and thick runners. Breaking the surface, a cluster of inflated pods, the first line of defense against submersion, formed the base of each individual plant. Rising out of the center of each cluster, several strong stalks sought the sky, and from each stalk six lavender petals fanned out, the topmost of which was marked in hues of blue and yellow. The seven generations, each comprised of several hundred plants, formed a dense mat about the size and shape of a supermarket aisle, hugging the shoreline and spilling out into the mouth of the stream that drained the marsh. The mother plant had arrived from upstream, via another small stream, when her runner was severed by a foraging animal. She traveled downstream, sustaining several wounded leaves and roots, finally settling in this placid pool, where she proceeded to reproduce with only herself. If there had been a male in the new pool, she could have reproduced sexually, producing seed which could last in the sediments, in case of flood or drought, for nearly thirty years.

One day, it is uncertain when or how, someone noticed how lovely and hardy the water hyacinth colony was; or perhaps, like the mother, one of the daughters floated downstream and was collected by a fisherman. Perhaps it was an amateur botanist, scouting exotic plants for the budding aquatic garden hobbyists in the US. He may have gathered several plants and seeds, transporting them by plane in tanks of water back home. If so, this may be how several live specimens came to the US in the 1880s, to a fair where they were bought by the hobbyists and bred for resale. From those few fledgling plants, many daughters were produced which spread from pond to pond, lake to lake, state to state, and nation to nation. In the absence of any natural predators, and with the encouragement of gardeners entranced by its beauty, the water hyacinth spread until it became naturalized throughout most of the southern US. It spreads so quickly and successfully that people became wary; anything that outpaces humans seems to come under scrutiny from biologists, state officials, and government agencies. Each of these parties discovered that, while water hyacinth is beautiful to look at, it is extremely hard to eradicate. Perhaps it spreads from a garden pond to a municipal sewer system; water hyacinth thrives in water polluted with excess nutrients. Then it may have clogged intake pipes to treatment plants, or perhaps multiplied aggressively in holding ponds. This scenario was replayed in thousands of similar places across the country; word began to spread through the grapevine that an exotic invasive had arrived and begun to spread. Exotic invasives thrive outside of their native habitats, because conditions are perfect for their growth and they have no natural predators in their new environments. They also outcompete native species, whose ecosystem functions people tend to disregard until they are gone. Water hyacinth accomplished all these deeds — the innocuous, lovely plant from South America crept into every corner of the southern US, and continues to spread, despite the fact it is illegal to buy, sell, or transport it; you face a hefty $500 fine and/or six months in jail if you do so. Despite the threat, it is still sold in nurseries, or traded privately–it is hard to resist something so attractive, and so easy to grow.

Water hyacinth is not just a problem in the US. In fact, it is considered to be the worst aquatic plant in the world; from India, Bangladesh, Pakistan, the Philippines, Thailand, Malaysia and Australasia to most of subtropical Africa it has spread ceaselessly. In Eastern Africa, where the problem seems to be particularly intense, water hyacinth is classified as a noxious weed, and its culture is illegal in some places. African lakes have a long history of being invaded — not by warring factions, but by plants. The lakes are usually large, and have many niches for different types of plants; colonists also had their own ideas of appropriate flora, and tended to import species to make their new environs feel more like home. Home, apparently, was full of water hyacinths. The lakes are also located in a subtropical climate, which worldwide is just behind the tropics in terms of biodiversity–the number of species which can comfortably call it home.

Lake Victoria provides a perfect case study for water hyacinth invasion. Unlike the early spread in the US, the Lake Victoria spread happened much more recently, and good records of the infestation were kept. It seems that water hyacinth first came down the Kagera River, which empties into Lake Victoria, in large floating mats dislodged from upstream. Some may also have come from ornamental ponds around Nairobi, Kenya as well; water hyacinth was kept in ponds at least as far back as 1957, and it made its first documented appearance in Lake Victoria in 1959. Water hyacinth is still flowing down the Kagera River and into Lake Victoria; one study estimates that approximately 0.2-1.5 hectares (ha) enter the lake daily. In addition, the study calculated 1% growth per day by already present plants, meaning that 64 ha along the shoreline could generate 0.64 ha/day of new growth in a single day. In the Congo River, another area where water hyacinth has made its home, two plants were observed to produce over 1200 daughters in the space of four months. This rapid rate of growth accounts for the rate of spread in Lake Victoria; in 1991, water hyacinth was observed to cover 1% of the 27,000 square miles of lake. The coverage peaked in 2001 at about 5% (17,000 ha) and decreased due to the implementation of some control methods. Recently it has begun to grow again and in [2], it is estimated to cover about 1.5% (5,000 ha) of the lake. The plant will continue to grow in the absence of controls: if two plants can produce 1200 daughters, and each of those 1200 daughters can produce at least 600 plants apiece in perpetuity, imagine how rapidly the rate of coverage will begin to increase! At various points in time, 80% of the Ugandan shoreline is infested to at least ten meters out, even beyond, and areas such as the Mwanza Gulf are sometimes blocked over their entire width–in this case, almost five kilometers Away from the shoreline, free-floating dense mats of water hyacinth nearly 600 ha in area sometimes are blown into bays on the North side of the lake by southerly winds, blocking traffic for days, until they are blown back out again by a north wind.

One of the most important factors accounting for the rapid spread of the water hyacinth is that it has evolved to reproduce and grow maximally in many conditions. The plant is self-pollinated, with the assistance of insects, but also can (and usually does) reproduce vegetatively, through runners like the ones described in the Brazilian plant above. If it can reproduce sexually, it will set seed, which sinks to the bottom, lodging in the sediments, and can last for up to 30 years–until conditions are right for germination. The plant can survive temperatures as high as 34°C, and as low as freezing; the leaves can be killed by frost, but in order to kill the entire plant and halt growth, the rhizome tips must be completely frozen. Estimates of the time it takes for a given number of plants to double range from 8-20 days, depending on environmental conditions. One study calculated the average growth rate of the plant to be 10-12 g per square meter per day, with a maximum of 45-50 g m^{-2} d^{-1}. Another put it in different terms; a single plant was observed to produce 140 million daughter plants per year, enough to cover 140 ha of water with a fresh weight of 28,000 tons of plant matter.

If water hyacinth were just another lovely plant, easy to keep when it was desired, simple to divest when it was no longer wanted, this chapter would not exist. Unfortunately, the plant has proved to be more terrible than Godzilla, and larger than King Kong. It is, to many who try to make a living from Lake Victoria and such similar places, a true monster, negatively affecting navigation, fisheries, industry, water supply, and health. The dense mats block bays and the movement of ships, limit landing sites, hamper docking of ferries, and damage the engine cooling systems of boats. At Port Bell, the turnaround time for ferries has increased from 6 to 12 hours since the occurrence of water hyacinth, and over 1,000 liters of fuel are sometimes used by a ferry just to break through the snarls. The Uganda Railways Corporation has at times been known to spend over $12,000 per week for mechanical removal of water hyacinth at Port Bell, and Kisumu Port authorities may be forced to do the same–the plant has brought port activity to a standstill, holding up food consignments and other shipments for weeks.

Fishermen, when asked about the problem, might point to the mats, shaking their heads, and tell a tale of once abundant species of fish that have declined rapidly since the invasion. The rapid growth of water hyacinth lowers oxygen levels at the shore (since respiration consumes oxygen), forcing fish to occupy less desirable habitat. It also reduces the availability of spawning grounds, though ironically the mats provide refuge for Nile Perch, another introduced species which is outcompeting native fish. At least one study has suggested that the hyacinth provides cover for the diminished populations of cichlid fish as well, providing the first example of potential benefit to the lake.

Activities such as industry are affected as well. Hydropower, an important source of energy for the surrounding nations, is severely interrupted when the intake screens and cooling filters of turbines are clogged. This results in higher costs for a less reliable source of energy, something the impoverished villages and cities surrounding Lake Victoria cannot easily afford. Some have estimated that 25 million people around the lake will potentially be affected by the plant, racking up economic losses of 150 million dollars annually from an estimated 5 percent reduction in lake quality. The reader is invited to speculate how one might measure "lake quality" in quantitative terms.

Local fishermen are thwarted by mats of vegetation blocking exist from their beaches and harbors to open water. A harbor or bay can be blocked overnight when the wind blows in a green iceberg. Commercial fishing boats face the same difficulty. Some even worry that actual changes in water level may result from the hyacinth, which has an unusually large transpiration rate. It functions as an efficient pump, moving water from the lake back into the atmosphere.

In light of the given information, it seems at first quite clear that something must be done to eradicate the water hyacinth, or at least to halt its spread out of its native habitat. This attitude is the traditional human response to change in habitat, whether involuntary (as in this case) or voluntary. In the US we can see this attitude displayed in the everyday behavior of people who move from the wet, green, grassy east coast to the mediterranean climate of California and then proceed to attempt to recreate their lawns and gardens as if the actual annual rainfall had not changed. Only a few reactionary holdouts find their dogged determination odd. It is this same aspect of human nature that explains why no one runs about saying, "We must adapt to the water hyacinth!" Instead, we compare the humble plant to Godzilla, labeling its completely natural behavior as "invasive," a negative term usually reserved for advancing armies of human beings.

In fact, one reason that the hyacinth can grow so rapidly in Lake Victoria is that it is very well fed by nutrients that run into the lake at a rate that is made all the larger for human activity. The algae bloom described in Chapter 3 is the forerunner of the water hyacinth.

26.2. The model

Agencies worldwide have attempted to address the issue, some more successfully than others, some still trying. Herbicides such as 2, 4-D and glyphosphate have been tried, though the effect has been minimal. It is difficult to reach all the plants with just one spraying, and multiple sprayings might negatively affect other aquatic life. Also, the mass of dead plants from the spraying proceed to decay *in situ*, releasing nutrients into the water and depriving it of oxygen, which could lead to eutrophication and fish kills. Additionally, there is no guarantee that reinvasion by sunken seed will not occur months, even years later.

Another possible option, still being avidly explored, is cutting and harvesting of the plant mechanically. Vaughn Co. USA manufactures a device that can harvest the plant without detaching the seeds; Aquarius Systems US won a World Bank contract to chop 1500 ha of water hyacinth in 12 months, using two "swamp devils" and a harvester. They plan to use the harvested biomass for fertilizers and crafts. There have been no significant results from this approach, especially in shallow areas where the infestation is worse, possibly since the process has been compared to cutting away a small island. The harvesting machines can keep commercial fisheries and transportation alive, however, by clearing channels from dockside to open water. The chopping machines are also used to manage floating islands of hyacinth that sail regularly into Kisumu Port at a speed of 3 to 4 knots.

If the chemical and mechanical methods fail, there is still one other option, and it is biological. In 1997, 35,000 weevil eggs (*Neochetina bruci, Neochetia eichhornia*) were released in selected areas. There are virulent pathogens which exist specifically to target water hyacinth, and if a native strain can be developed, it will be tried. It has been suggested that the fungus *Cercospora rodmanii* may improve the success of the beetles, and that control may even be achieved by manatees, as it has in Guyana, snails, which have worked to some extent in the US, and mites (*Tetranychus telarius*), used in Belgium. The main theme seems to be that water hyacinth is very determined, but its human foes are even more determined to eradicate it; perhaps, as some have suggested, the best solution is every solution, integrated and managed intelligently. A mathematical model could prove extremely useful in guiding this process.

Our previous models basically described the relationship between two species, which could be diagrammed as boxes with a connecting line standing for the relationship between two boxes. Each box represents a quantity that can change. Any change in a given quantity should be written in terms of the boxes connected to that quantity. In the case of Lotka Volterra or competition between species, we would have two boxes, and the relationship between them was represented by the mixed term in the differential equation. When we erased that term (the product of the two populations) we got an equation for the growth of one population in the absence of the other one.

Any model describing the effects of the water hyacinth will have a lot of boxes. The hyacinth, the weevils, the perch and cichlid are all interconnected. If we believe that the cichlid benefits from hiding in hyacinth, then the equation for change in cichlid population should include a term representing the advantage offered by the vegetation. The beetle eats hyacinth and is therefore dependent on its population, but we have to decide if the Lotka Volterra term applies in this situation of stationary prey. If not, we will have to invent another kind of term that represents the interaction better. If we include the effects of harvesting, then we have to decide what likely human behavior might be. Will it be a constant amount taken each year or will the amount vary according to some rule? If we want to take the general degradation in oxygen into account, we have to decide which species that affects and according to what rule. We have many boxes, or "compartments" in our model and each will have its own differential equation describing its rate of change, built out of its natural growth rate and all the relationships described in the diagram we build. Fortunately, once we have our equations, any decent numerical program will be able to produce solutions for us, either as time plots or as projections of the phase portrait onto any two variables we choose. See Figure 26.1 for one possible configuration of boxes.

The creators of this configuration had four differential equations. For the hyacinth population, as an example, it had:

$$\frac{dw}{dt} = gw - few - k.$$

The first term is the natural growth rate of the hyacinth, (w). The second term is a standard Lotka Volterra interaction with the beetle population, (e). Notice the extra term at the end. These authors used this term to model human harvesting.

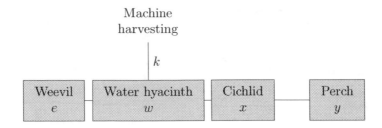

FIGURE 26.1. Proposed box model for hyacinth/beetle/cichlid/perch.

They are saying that the amount harvested is constant. Notice the assumptions that go with each of these terms.

The first term is saying that, in the absence of harvesting or beetles the hyacinth grows exponentially. It is the same equation we saw with algae in Chapter 3. Of course, there is a limiting factor to the growth of hyacinth, namely the size of the lake. If one percent of the lake is currently covered, are we still far enough from full to ignore the carrying capacity of the lake? If not, we should be using the logistic equation for this term.

The second term is saying that the predator prey relationship is a sort of hide-and-seek kind of relationship like that of fish populations. In fact, the hyacinth is fairly stationary. Perhaps, as the beetle population increases, the beetles do have to search for suitable unused spots to lay eggs, so this is a reasonable description of their activity.

The last term is claiming that human harvesting behavior is constant. Is this what people actually do? Presumably a more accurate model of human behavior could be invented.

The authors of the above model had equations for all the boxes in their diagram.

In this example there is no harvesting and we see the classic periodicity associated to Lotka Volterra. The weevil does not eradicate the hyacinth and the hyacinth has recurrent periods of high population every five years or so.

Here we see the hyacinth completely controlled and the beetle population going to zero or nearly so. We must stress that, in this example, the harvesting behavior is so extreme as to be impossible for practical purposes. One of the very difficult problems for this model is deciding what constants are reasonable based on the literature. The authors of this particular model concluded that harvesters were unnecessary in the long run because of the long term behavior of their model, but failed to analyze whether the recurrent blooms of hyacinth were large enough to be troublesome for human activity. They also failed to notice that overharvesting, followed by a cessation of harvesting, results in larger oscillations in the Lotka Volterra model. Pushing one population toward zero lands you on one of the outside loops of the phase portrait, resulting in a time series whose oscillations are farther from the equilibrium point and thus very large. We leave it to the reader to develop a better model, more delicately analyzed.

It is worth mentioning at this point that the model in this chapter was drawn from student work, and the equations and outputs are entirely due to these students working from this text and the literature. By far the hardest part of their work was figuring out what the values to give to the relevant parameters. When measuring the growth rate of hyacinth, choices of unit vary from square meters of coverage to weight, making the task of unit conversion and consistency a tough part of the problem.

In the interests of inquiry and fairness, let us forget about eradicating the poor plant and ask instead what adaptations humans might make during a periodic hyacinth boom. Surely a plant this fecund must have its uses and, if we can find them, we will surely be rewarded. It should not surprise us that many people have found ways to capitalize on the economic uses of this plant.

The harvested product has been used for silage and soil compost in Uganda, as well as for weaving twine and rope (Thai Trading company offers water hyacinth cord on its website). There are two biogas plants which make use of it as fuel, and on a small scale it has even been used for livestock feed. It is not useful freshly harvested as fodder, since it is prickly and the seed can be spread in feces, but this is not a new dilemma. Farmers in Southeast Asia have developed a successful recipe for the feed, which calls for it to be boiled to a paste, mixed with bran, maize, and salt, and consumed within three days lest it go foul. A more specific recipe requires 40 kg water hyacinth, 15 kg rice bran, 2.5 kg fish meal, and 5 kg coconut meal. As a fuel, one kg of dried water hyacinth will produce 474 liters of biogas with 75% methane; perhaps Eastern Africa will invent the safe and cheap natural gas-powered car before the West does!

Industrially, water hyacinth has been successfully used to produce cellulose, which is a sacharifying agent used to produce fuel and chemicals. As a bonus, it is much less expensive than cotton or solka floc, the other two substances used as agents. One byproduct of the process is Xylanase, which is used in the paper industry, to upgrade animal feed, and for the clarification of fruit juice and wine. Another byproduct is a potent plant pathogen, and still another is fertilizer and soil conditioners. Best of all, this process has been established successfully on a commercial scale, albeit in India.

Still one more use involves water treatment; it turns out water hyacinth has quite an affinity for heavy metal ions, such as gold and lead, which contaminate our groundwater. The San Pasqual Aquatic Treatment Facility, located in San Diego County, California, uses water hyacinth with reverse osmosis and chemicals to treat over 1,200,000 gallons of sewage per day, using the gray water byproduct to water nearby lawns. The plant is ideal for this use because it quickly takes up phosphorus and nitrogen, two of our major aquatic pollutants, it outcompetes phytoplankton, which cause fish kills, and metabolizes organic substances such as phenol, improving water transparency.

In fact, people have found water hyacinth so useful that a library search will turn up articles describing mathematical models for nutrient usage of hyacinth, with the goal of finding a fertilization regime that actually maximizes its growth rate. And on the shores of Lake Victoria we find small scale businesses arising that collect and process water hyacinth. It is made into woven furniture that is of high quality, suitable for trade and export. In some provinces, at some times, the collection of water hyacinth without a permit has been banned. This type of small scale, geographically determined harvesting could also be accounted for in a model.

From local village tribunals and international task forces surrounding the lake to the pockets of multinational agencies such as the International Monetary Fund and the World Bank, it seems everybody has a stake in the future of the Lake

Victoria economy, and in preventing the spread to other nations of the life-loving water hyacinth. Nobody has yet achieved a successful method of eradication, or perfected a commercial scale safe harvesting method; but everybody seems to be working on it. As of 1999, the World Bank and the Global Environmental Facility had loaned 23 million Kenya shillings ($1US = 5 shillings); others had offered to solicit $2.5 million more for mechanical removal. There was money available, and now there needed to be a plan. Eradicate it? Use it? Both? It is up to the nations whose wealth and livelihood depends on the Lake Victoria economy, the regional and local governments and people who will ultimately be affected.

Meanwhile, nature continued to adapt. Aquarius Systems was kind enough to share with us a report from a demonstration project in Kenya. Part of the report describes the state of the hyacinth at the time chopping started in October 1999, several years after the introduction of the beetle to the lake. Evidently the beetles have considerably weakened the hyacinth, but as Lotka Volterra predicts, failed to eradicate it. Masses of floating hyacinth became colonized by secondary growth consisting of at least 23 plant species, much of which was Hippo Grass and papyrus. The personnel from this project estimated that about 30 percent of the visible plant material was hyacinth and 70 percent was other vegetation. Bushes and trees were identified on some of the floating islands and on others the weeds were more than 10 feet high. No one knows whether this secondary colonization would have happened without the introduction of beetles. It seems possible now that, for hyacinth to be used as an industrial base, the weevils themselves will have to be controlled, at least in some locations. We close this chapter with a quote from the Aquarius Systems report:

> What we can say for sure is had we not been chopping in the Port of Kisumu this winter and keeping the floating islands that arrived each day in check, it would have looked like a forest and been unusable in a matter of months.

For additional information from a scientific point of view about the Water Hyacinth and the Weevil, see [8, 6, 7, 5, 4, 3, 1]. For a weevil Hyacinth model see [6].

26.3. Projects

(1) Design a box model that reflects your understanding of the water hyacinth problem in Lake Victoria. Write down equations for the rates of change for each of the quantities in your model. Experiment with your model to see what it predicts as long term behavior of the beetle and hyacinth.

Modify your model to take into account the effect of human harvesting. What is your recommendation to the governments of Tanzania, Uganda, and Kenya? Can they safely discontinue harvesting water hyacinth of the beetles are present? Should they keep the harvesting machines? Write your answer as a technical report to a hypothetical committee appointed to study this problem.

(2) The effect of hyacinth on both cichlids and Nile perch fry is to provide cover, protecting them from predators. Make a new box model that attempts to model this effect for these fish. What does your model predict about perch and cichlids? Is there a benefit for the cichlids?

(3) One hypothesis found in the literature is that floating mats of hyacinth have the capacity to reintroduce populations of cichlid into areas where they may have become extinct. Build a model that tries to capture this phenomenon.

As in the case with competing species, different initial conditions sometimes lead to radically different long-term outcomes. Experiment with your model to see if the long-term outcome is consistent, independent of initial conditions.

(4) An alert reader of the last chapter might be wondering if the assumptions that lead to the usual predator prey equations might be inappropriate for a situation where the prey is a plant. After all, plants don't run away. They behave more like a renewable resource than an elusive quarry. In evolutionary response, the water hyacinth weevil doesn't need to spend a lot of personal resources searching for its prey and thus might not be subject to the same dynamic as the predator prey equation predicts. In particular, suppose the weevil and hyacinth obeyed the standard (damped) equations:

$$N = \text{hyacinth}$$
$$W = \text{weevil}$$
$$N' = rN(1 - N) - cNW$$
$$W' = -gW + cNW.$$

Then the second equation, $W' = -gW + cNW$, says that at low concentrations of hyacinth, the weevil is dying off. That is, if cNW is less than gW, the growth rate W' is negative. Perhaps this is not a good assumption, in that only a small amount of hyacinth might be necessary to accommodate the reproduction needs of weevils.

This system might also be subject to a nonlinear predator functional response such as the one we saw in Chapter 20. Such a response might further be supposed to have a large effect on the hyacinth population but less of an effect on the weevil population. That is, the death rate of hyacinth would be very much affected by the rate of consumption by weevils, but the birth rate of weevils might actually depend more on the amount of hyacinth available rather than the rate at which it is eaten. This situation would occur if weevils quickly achieve their maximal feeding rate.

Such assumptions are built into a model studied by John Ross Wilson [6, 7]. This model replaces the equation for weevil growth by a modified logistic equation whose carrying capacity is given as a proportion of the water hyacinth density. Its equation for plant growth includes a nonlinear predator response like the one in Chapter 20 of this text. Furthermore the model includes a slight steady influx of hyacinth into the system, which we set to zero for the purposes of the above. The units are all in terms of water hyacinth and weevil densities (grams of either hyacinth or weevil per square meter). The model looks like this, with N representing hyacinth and W the weevil:

$$N' = rN(1 - N/k) - c(N/(N + h))W$$
$$W' = sW(1 - jW/N).$$

Wilson's model has three advantages over the standard predator prey model we explored above. First of all, its assumptions seem to fit the situation better. Second, most of the constants that appear in these equations have been measured to some extent. Third, for the purposes of this text, Wilson's model exhibits some very interesting behavior as its parameters are varied. The kinds of behaviors we see in this model are actually quite typical of complex systems, and every modeler needs to be aware of the possibility that these might occur.

(a) In [6, 7] some parameters are given values: $r = 0.08$, $c = 4$, $h = 200$, $s = 0.06$, $j = 50$. What do the constants r, c, h, s and j represent? What are their units?

(b) What happens to the system for large N?

(c) What happens to the system for small N or large W?

(d) The value for h is the most difficult to determine experimentally. Why is that? What happens if we change the value of h. Let h be 2000, 700, 550 and 200? Plot the solutions to the system for each value of h. What do you notice?

(e) Plot the phase portraits for each value of h. What do you notice?

(f) You should have seen limit cycles and Hopf bifurcations. Look up what these mean and restate your observations from above in these terms.

(g) Will the weevil be effective in controlling the water hyacinth?

Bibliography

[1] Thomas P Albright, TG Moorhouse, and TJ McNabb. The rise and fall of water hyacinth in lake victoria and the kagera river basin, 1989-2001. *Journal of Aquatic Plant Management*, 42(JUL.):73–84, 2004.

[2] Peninah Aloo et al. A review of the impacts of invasive aquatic weeds on the bio-diversity of some tropical water bodies with special reference to lake victoria (kenya). *Biodivers. J.*, 4:471–482, 2013.

[3] Ted D Center, F Allen Dray Jr, Greg P Jubinsky, and Michael J Grodowitz. Biological control of water hyacinth under conditions of maintenance management: can herbicides and insects be integrated? *Environmental Management*, 23(2):241–256, 1999.

[4] David A Cornwell, John Zoltek Jr, C Dean Patrinely, Thomas deS Furman, and Jung I Kim. Nutrient removal by water hyacinths. *Journal (Water Pollution Control Federation)*, pages 57–65, 1977.

[5] Tim A Heard and Shaun L Winterton. Interactions between nutrient status and weevil herbivory in the biological control of water hyacinth. *Journal of Applied Ecology*, 37(1):117–127, 2000.

[6] John R Wilson, Niels Holst, and Mark Rees. Determinants and patterns of population growth in water hyacinth. *Aquatic Botany*, 81(1):51–67, 2005.

[7] John RU Wilson et al. The decline of water hyacinth on Lake Victoria was due to biological control by *Neochetina spp. Aquatic Botany*, 87(1):90–93, 2007.

[8] Gui-Ying Xie and Jin-Chun Guo. Water hyacinth occurrence, control, and utilization. *Pesticides*, 10:004, 2005.

Part 5

Infectious disease modeling

CHAPTER 27

SIR Model for Infectious Diseases

The kind of modeling we have been doing has been applied to the study of infectious disease since the early 1900s [2]. So-called "Compartmental models in epidemiology" are quite common and have proven quite useful in helping different kinds of researchers (mathematicians, public health researchers, etc.) understand the complex dynamics of infectious diseases. Much of what follows in this section is based on [4].

27.1. Biological context

In late 2009 a cholera epidemic swept across Kenya. In that year there were 11,425 reported cases, including 264 deaths [3]. Cholera, called *kipindupindu* in Kiswahili, is an acute diarrheal illness caused by a bacterial infection in the intestine. People exposed to cholera get sick quickly (within an hour say) and those who do not get treatment can die within a day. It spreads quickly and easily: taking a sip of water from a cup used by a person with cholera could be enough for the illness to spread to another person.

About 100 million bacteria must typically be ingested to cause cholera in a normal healthy adult. Children are more susceptible, with two- to four-year-olds having the highest rates of infection. Persons with lowered immunity, such as persons with AIDS or children who are malnourished, are more likely to experience a severe case if they become infected. We point out that during the epidemic of 2009 both of these assumptions were relevant: the country was suffering from drought and because sub-Saharan Africa in general, and Kenya, particular, were struggling to control the spread of HIV/AIDS.

There is no cure for cholera, but it can be treated if it is caught in time. The typical treatment is oral rehydration therapy–the replacement of fluids with slightly sweet and salty solutions. In severe cases, intravenous fluids, such as Ringer's lactate, may be required, and antibiotics may be beneficial. Testing to see which antibiotic the cholera is susceptible to can help guide the choice of which strategy to use.

Cholera is now no longer considered a pressing health threat in Europe and North America due to filtering and chlorination of water supplies, but still affects populations in developing countries like Kenya. Roughly 3-5 million are infected each year and around 100,000 deaths are attributed to cholera each year. There is a vaccine for cholera and it provides about six months of reasonably good protection from the disease. Additionally, in some countries (e.g. Australia), people suffering from cholera are required to be quarantined.

Because of the scope of the cholera pandemic, it would be useful to be able to model how the disease spreads in a population. To do this, we divide the population into three groups: the susceptible (S), the infected (I) and the recovered (R). A model based on this kind of division is called an SIR model. Such models have been used to study many different kinds of infectious diseases. Instead of considering S, I, and R (the raw number of susceptibles, infecteds and recovereds[1]), we often consider the fractions of the total population, assumed to be some constant N. I.e., we define $s(t) = S(t)/N$, $i(t) = I(t)/N$ and $r(t) = R(t)/N$, where N is the total population, and we model those, instead.

The questions we want to consider are:

- Will there be an outbreak? I.e., will i get increase?
- Can we reduce the likelihood of an outbreak by reducing the initial size of s via vaccination?
- Can we reduce the likelihood of an outbreak by reducing the transitions from s to i by quarantining?

The answers to the last two questions are clearly both "Yes", but we will be interested in seeing how those phenomena are manifested in the models.

27.2. The model

We first describe the assumptions of the model.

We ignore births, deaths and immigration and so no one is added to the total population. So, in particular, we have

$$S(t) + I(t) + R(t) = \text{constant total population}$$

for all time t.

EXERCISE 27.1. From this conclude that

$$\frac{dS}{dt} + \frac{dI}{dt} + \frac{dR}{dt} = 0.$$

The only way a person leaves the susceptibles is if they become infected. We therefore assume that the rate of change of $S(t)$ (or $s(t)$, depends only on the number of susceptibles, the number of infecteds and the amount of contact between infecteds and susceptibles. In particular, suppose that each infected has a fixed number b of infectious contacts per day and that instantaneous and uniform mixing of the two populations occurs. This leads to a differential equation:

$$\frac{dS}{dt} = -b \times s \times I.$$

This should remind the reader of part of the Lotka-Volterra model.

EXERCISE 27.2. Explain where the minus sign comes from.

EXERCISE 27.3. What are the units of b? Are they consistent with the verbal description above?

EXERCISE 27.4. Why is it also true that

$$\frac{ds}{dt} = -b \times s \times i?$$

---

[1]We understand that pluralizing these words is not correct grammatically, but the use of these words is rather standard in the field.

In addition to our assumptions on how people transition from a susceptible to an infected, we also need to make an assumption about how people transition from being an infected to a recovered. We assume that a fixed fraction k of the infecteds during any given day. In our case, cholera generally lasts a week and so, or average, each day about one seventh of the infecteds become recovereds. This leads to

$$\frac{dr}{dt} = k \times i,$$

where, for cholera, $k = \frac{1}{7}$.

EXERCISE 27.5. What are the units of k? Are the units consistent with the verbal description above?

Based on these three assumptions we get the complete system for our model of cholera:

$$\frac{ds}{dt} = -b \times s \times i$$
$$\frac{di}{dt} - b \times s \times i - k \times i$$
$$\frac{dr}{dt} = k \times i.$$

Of course, to use this model, we would have to know some initial conditions.

EXERCISE 27.6. Justify the equation for $\frac{di}{dt}$. Hint: Use Exercise 27.1.

EXERCISE 27.7. What if $i(0) = 0$? What happens?

EXERCISE 27.8. Based on the equations and verbal descriptions above, draw a box model for cholera.

EXERCISE 27.9. Can you see how a low peak level of infection can nevertheless lead to more than half the population getting sick? Explain.

27.3. Analysis of the model

While we have reason to guess that $k = \frac{1}{7}$, we do not yet have a value for b. In Figure 27.1, we see how the graph of $i(t)$ changes by changing b.

EXERCISE 27.10. Is Figure 27.1 consistent with the physical meaning of the constant b?

EXERCISE 27.11. Observe that in the graph corresponding to $b = 0.12$ we see that $i(t)$ is always decreasing. By considering the factored expression

$$\frac{di}{dt} = bsi - \frac{i}{7}$$
$$= i \left(bs - \frac{1}{7} \right),$$

could you have predicted this? Hint: recall $i(t)$ is decreasing if and only if its derivative is negative.

We have guessed that $k = \frac{1}{7}$ for cholera, but let's see how changing k affects the graph of $i(t)$ and check to see if it is consistent with our understanding of the model. Additionally, it may be that the value of k is different in different populations (e.g., maybe k is different for children than it is for adults).

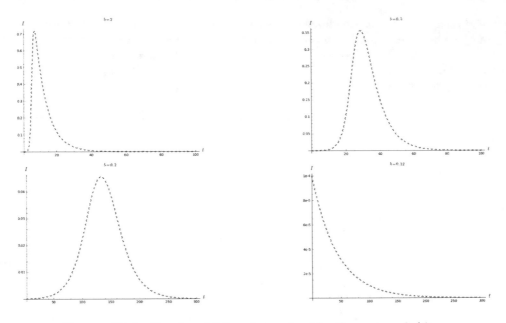

FIGURE 27.1. Graphs of $i(t)$ as determined by the system $ds/dt = -bsi$, $di/dt = bsi - ki$, $dr/dt = ki$ for various values of b and with k fixed at $\frac{1}{7}$. Notice that the vertical scale of each graph is quite different.

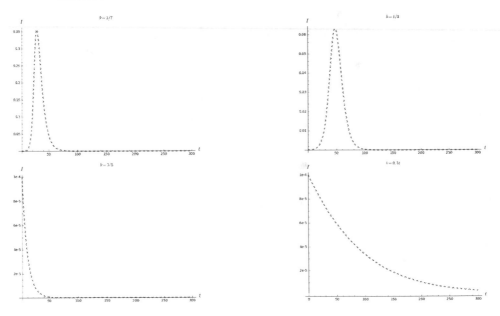

FIGURE 27.2. Graphs of $i(t)$ as determined by the system $ds/dt = -bsi$, $di/dt = bsi - ki$, $dr/dt = ki$ for various values of k and with b fixed at $\frac{1}{2}$. Notice that the vertical scale of each graph is quite different.

EXERCISE 27.12. Are the images in Figure 27.2 consistent with our understanding of what k means physically?

EXERCISE 27.13. It appears that for k just larger than 0.5, the qualitative behavior of $i(t)$ changes drastically: in particular, it is always decreasing. Explain why that is. Hint: it may be useful to observe that $s(t)$ can be 1 at its largest.

27.4. Implications of the model

While there is a direct way to compute k (it is the reciprocal of how long a typical infection lasts), there is no direct way to compute b. We now describe an indirect way to do this.

Consider

$$\frac{b}{k} = b \times \frac{1}{k}$$

= number of close contacts per day between an infected and a susceptible

$\times$ length of infection

= number of close contacts between an infected and a susceptible.

We call this ratio the *contact number* and call it c.

EXERCISE 27.14. Suppose you knew k and could calculate c. How would you find b?

We now discuss a method to compute the contact number c. Once again, we consider our system:

$$\frac{ds}{dt} = -bis$$
$$\frac{di}{dt} = bis - ki.$$

Notice that if we could divide the second equation by the first, we would get something much simpler:

$$\frac{di}{ds} = \frac{\frac{di}{dt}}{\frac{ds}{dt}} = \frac{bis - ki}{-bis} = -1 - \frac{k}{bs} = -1 - \frac{1}{cs}.$$

EXERCISE 27.15. Explain why the Chain Rule makes the first step legitimate.

Now, as a function of s, then, we can write

$$i(s) = -s - \frac{1}{c} \ln(s) + A$$

where A is some unknown constant. We observe that this expression is independent of time.

EXERCISE 27.16. Why is this expression independent of time? Note: It is *not* because there is no explicit mention of time in the expression.

Now $s(0) = 1$ and $i(0) \approx 0$. If we assume that $i(\infty) \approx 0$, then $s(\infty) = s_\infty$, some equilibrium level of susceptibles. Using this we can deduce

$$-s(0) - \tfrac{1}{c} \ln(s(0)) + A = -s(\infty) - \tfrac{1}{c} \ln(s(\infty)) + A$$
$$\Rightarrow -1 - \tfrac{1}{c} \ln(1) + A = -s_\infty - \tfrac{1}{c} \ln(s_\infty) + A$$
$$\Rightarrow -1 - 0 + A = -s_\infty - \tfrac{1}{c} \ln(s_\infty) + A$$
$$\Rightarrow s_\infty - 1 = -\frac{\ln(s_\infty)}{c}$$

$$\Rightarrow c = -\frac{\ln(s_\infty)}{s_\infty - 1}.$$

So, if we could find s_∞, we would be able to find c (and therefore b). Luckily, s_∞ is something that can be found empirically. After an outbreak of cholera, for example, it would be possible to find what fraction of people were exposed to cholera but were not infected. This fraction corresponds exactly to s_∞.

One way to avoid an outbreak is to vaccinate part of the population. It is impossible to vaccinate an entire population (e.g., someone should have the right to refuse a vaccination due to religious preference or because they are immuno-suppressed) and so the question is, how much of the population needs to be immunized to avoid an outbreak? In other words, how big should $s(0)$ be in order to avoid an outbreak? (The expression $s(0) = 1$ means no one received a vaccine and $s(0) = 0$ means everyone received a vaccine.)

Now, note that if $\frac{di}{dt}$ were always negative, we would not get an outbreak. We consider

$$\frac{di}{dt} = bsi - ki = i(bs - k)$$

and note that $\frac{di}{dt} < 0$ if and only if $bs < k$. Since $\frac{ds}{dt}$ is always negative, we see that $s(0)$ is the maximum value that s can take on. So if $s(0) < \frac{k}{b} = \frac{1}{c}$, we will have $\frac{di}{dt}$ always negative and therefore $i(t)$ always decreasing.

27.5. Problems

PROBLEM 27.17. Another way to avoid an outbreak is to quarantine. Explain how you might incorporate quarantining a patient into the model. Run the model a few times to see what effect your tweak to the model has.

PROBLEM 27.18. From [4]. From 1912 to 1928, the contact number for measles in the US was 12.8. If we assume that c is still 12.8 and that inoculation is 100% effective–everyone inoculated obtains immunity from the disease–what fraction of the population must be inoculated to prevent a measles epidemic? Suppose the vaccine is only 95% effective. What fraction of the population would have to be inoculated to prevent a measles epidemic?

PROBLEM 27.19. An SIR model is just one of many compartmental models used in epidemiology. An SEIR (Susceptible-Exposed-Infected-Recovered) model is one that adds a compartment for people who have been exposed to the disease but are not infected. Write down a system of differential equations that could model such a disease and understand what role each of the constants in your model plays (i.e., what happens as you fix all but one and let the others vary).

PROBLEM 27.20. In some settings (e.g., Ebola [1]), it can be shown that the transmission rate b is constant until some intervention at time τ, at which point the transmission rate drops exponentially. Incorporate this tweak into the SIR model and discuss what happens qualitatively. In particular, consider how small (or large) τ can be to avoid an outbreak and what role the rate of exponential decay has on whether or not an outbreak can be avoided.

Bibliography

[1] Christian L Althaus. Estimating the reproduction number of ebola virus (ebov) during the 2014 outbreak in west africa. *PLOS Currents Outbreaks*, 2014.

[2] William O Kermack and Anderson G McKendrick. A contribution to the mathematical theory of epidemics. *Proceedings of the Royal Society of London A: mathematical, physical and engineering sciences*, 115(772):700–721, 1927.

[3] Anagha Loharikar et al. A national cholera epidemic with high case fatality rates, Kenya 2009. *Journal of Infectious Diseases*, 208(suppl 1):S69–S77, 2013.

[4] David A Smith and Lawrence C Moore. *Calculus: Modeling and application.* DC Heath & Company, 1996.

CHAPTER 28

Malaria

Diseases present classes of interesting problems to the modeler. There are basically three kinds of questions to consider: epidemiology, physiology and pharmacology. In the previous chapter we considered questions of epidemiology when developed a simple SIR model for cholera. Of course these questions are related. Understanding the physiology of a disease leads to better understanding of how to prevent transmission of it, which informs epidemiology models. Treatment of disease (including pharmacology) also depends on understanding its physiology (pharmacokinetics, an aspect of pharmacology, was covered in Part 3). But, as models, these three kinds of investigations are usually handled separately.

To model the physiology of an infection we have to understand what kind of parasite is involved, what its life cycle inside the body looks like, where it lives and what it damages, where it can hide. Some diseases, such as trypanosomiasis, have separate stages that differ in what organs of the body are affected, how severe the affliction is during that stage, the death rate, and whether a particular drug or other intervention will work. Some diseases have characteristic symptoms that should be predicted by an accurate model. A model that incorporates the correct life cycle of a parasite during an infection should be able to predict symptoms characteristic of that disease, such as the periodic temperature spikes characteristic of malaria infection. A model that correctly predicts the progress of the disease confirms our understanding of its physiology, while a model that expresses well-understood physiology mathematically is useful for suggesting interventions and epidemiology strategies.

> The prescription employed is — resin of jalap, and calomel, of each eight grains; quinine and rhubarb, of each four grains; mix well together, and when required make into pills with spirit of cardamoms: dose from ten to twenty grains. The violent headache, pains in the back, etc., are all relieved in from four to six hours; and with the operation of the medicine there is an enormous discharge of black bile, — the patient frequently calls it blood. If the operation is delayed, a dessert-spoonful of salts promotes the action. Quinine is then given till the ears ring, etc. Those who may try the remedy will do well to remember that the above doses are for strong adults.

–From a letter to James McWilliam from David Livingstone, 1860, on fever in the Zambesi

Pharmacology concerns the invention and implementation of drug therapy for a disease. Pharmacokinetics is the branch of pharmacology that looks at how drugs pass through the body–how they are absorbed, metabolized and eliminated. These processes are all described by models, which give a basis for assigning dosages and frequencies for any given medication. Some of the tropical infections require strong medications that have serious side effects. Keeping the concentration of drug in the bloodstream at an effective but non-toxic level is an important mathematical problem.

Epidemiology is the study of how a disease propagates throughout a population. The entire field makes extensive use of mathematical and statistical models to predict in advance whether a particular intervention will be effective. "Effective" in this case could mean different things, such as reduction of hospital caseload or the actual stopping of an epidemic. Models must take into account the methods of transmission of the disease, including to and from any animal vectors or reservoirs. Tropical diseases feature a lot of insect vectors, including flies, mosquitoes, and snails. Often the life cycles of these insects must be included in a model in order to answer basic epidemiology questions. Sometimes the most effective intervention is made on the insect population rather than the human population. Sometimes the cost of an intervention is built into the model as an additional point of comparison, and a particularly relevant one for poorer areas. Only through modeling can various proposed strategies be compared with one another before implementing them.

> October 28th.–At a quarter past five o'clock in the morning my dear friend Mr. Alexander Anderson died after a sickness of four months. I feel much inclined to speak of his merits; but as his worth was known only to a few friends, I will rather cherish his memory in silence, and imitate his cool and steady conduct, than weary my friends with a panegyric in which they cannot be supposed to join. I shall only observe that no event which took place during the journey, ever threw the smallest gloom over my mind, till I laid Mr. Anderson in the grave. I then felt myself, as if left a second time lonely and friendless amidst the wilds of Africa.

> –From the Journal of Mungo Park

Would Mungo Park have undertaken his expedition if he had known that 40 out of his 44 men would die? If the World Health Organization succeeds in putting a mosquito net over every African child at night, what will this do to reduce the death rate of malaria? Modeling has value at every point in our consideration of a disease, from our initial understanding of its physiology to high-level policy decisions about its control. In this chapter we study one of the oldest mathematical models of the kind we have been studying–the models for malaria.

28.1. Biological context

Malaria is a mosquito-borne infectious disease affecting humans and other animals caused by parasitic protozoans belonging to the *Plasmodium* type. The burden that malaria places on the health, economy and society of (mostly) tropical countries is enormous.

The cost of malaria is manifold. First, we consider the symptoms of the disease. Malaria causes symptoms that typically include fever, fatigue, vomiting, and headaches. In severe cases it can cause yellow skin, seizures, coma, or death. Symptoms usually begin ten to fifteen days after being bitten. If not properly treated, people may have recurrences of the disease months later. In those who have recently survived an infection, reinfection usually causes milder symptoms. This partial resistance disappears if the person has no continuing exposure to malaria.

Second, we consider the social cost. The disease is widespread in the tropical and subtropical regions that exist in a broad band around the equator. This includes much of Sub-Saharan Africa, Asia, and Latin America. In 2015, there were 214 million cases of malaria worldwide resulting in an estimated 438,000 deaths, 90% of which occurred in Africa. Rates of disease have decreased from 2000 to 2015 by 37%, but increased from 2014 during which there were 198 million cases. The disruption this disease causes to families and communities is enormous.

Finally, we consider the economic cost. Malaria is commonly associated with poverty and has a major negative effect on economic development. In Africa, it is estimated to result in losses of US\$12 billion a year due to increased healthcare costs, lost ability to work, and negative effects on tourism. See [1] for an example of combining economic models and malaria models to illuminate a pathway out of poverty for communities stricken by malaria.

Malaria is typically diagnosed by the microscopic examination of blood using blood films, or with antigen-based rapid diagnostic tests. Methods that use the polymerase chain reaction to detect the parasite's DNA have been developed, but are not widely used in areas where malaria is common due to their cost and complexity.

The risk of disease can be reduced by preventing mosquito bites through the use of mosquito nets and insect repellents, or with mosquito control measures such as spraying insecticides and draining standing water (see [6], for example). Several medications are available to prevent malaria in travellers to areas where the disease is common. Occasional doses of the combination medication sulfadoxine/pyrimethamine are recommended in infants and after the first trimester of pregnancy in areas with high rates of malaria. Despite a need, no effective vaccine exists, although efforts to develop one are ongoing. The recommended treatment for malaria is a combination of antimalarial medications that includes an artemisinin. The second medication may be either mefloquine, lumefantrine, or sulfadoxine/pyrimethamine. Quinine along with doxycycline may be used if an artemisinin is not available. It is recommended that in areas where the disease is common, malaria is confirmed if possible before treatment is started due to concerns of increasing drug resistance. Resistance among the parasites has developed to several antimalarial medications.

28.2. The model

The dynamics of the spread of malaria have been studied mathematically for more than a century and, as such, there are many malaria models to be found in the literature. We recommend [4] for a high-level survey of the models and [9] for a more quantitative survey of some of the models. We focus on the Ross-Macdonald model originally due to Ross [7] and improved by Macdonald [3].

The model we describe takes into account the disease in both humans and in female *Anopheles* mosquitoes. In humans, the parasites grow and multiply first in the liver cells and then in the red cells of the blood. In the blood, successive broods of parasites grow inside the red cells and destroy them, releasing daughter parasites that continue the cycle by invading other red cells.

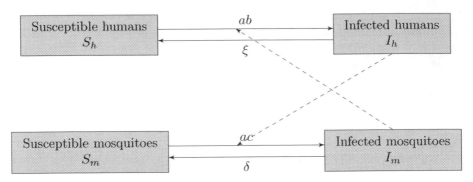

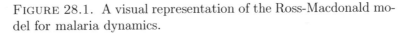

FIGURE 28.1. A visual representation of the Ross-Macdonald model for malaria dynamics.

The blood stage parasites are those that cause the symptoms of malaria. When
certain forms of blood stage parasites are picked up by a female *Anopheles* mosquito
during a blood meal, they start another, different cycle of growth and multiplication
in the mosquito.

After 10-18 days, the parasites (as "sporozoites") are found in the mosquito's
salivary glands. When the *Anopheles* mosquito takes a blood meal on another
human, the sporozoites are injected with the mosquito's saliva and start another
human infection when they parasitize the liver cells.

Thus the mosquito carries the disease from one human to another. Unlike
for the human host, the mosquito vector does not suffer from the presence of the
parasites.

In Figure 28.1 we see a visual representation of the model.

The assumptions that we make are similar to the ones we made in Chapter 27–
this is perhaps not surprising since what we have here are two coupled SIS models.
In particular, we assume

- the number of mosquitoes M is constant, as is the number of humans H;
- mosquitoes are either susceptible or infectious, as are humans;
- the rate at which mosquitoes bite humans is proportional to the number of
  mosquitoes but independent of the number of people (this is a functional
  response as in predator-prey modeling);
- populations mix uniformly and immediately;
- the populations of both are large;
- we ignore incubation periods (this would add an Exposed compartment).

Now we can give the algebraic description of the models:

$$I'_h = a \times b \times I \times \left(\frac{H - I_h}{H}\right) - \xi \times I_h$$

$$I'_m = a \times c \times (M - I_m) \times \frac{I_h}{H} - \delta \times I_m.$$

The parameters are as follows: a is the mosquito biting rate; b is the mosquito to
human transmission probability (per bite); c is the human to mosquito transmission
probability (per bite); ξ is the human recovery rate (average duration of human
infection, D_H, is $\frac{1}{\xi}$); δ is the mosquito death rate: average duration of mosquito
infection, D_M, is $\frac{1}{\delta}$.

EXERCISE 28.1. Recall that one of the assumptions is that H and M are both constant. Use this fact to find differential equations to model S_h and S_m.

28.3. Implications of the model

Recall Chapter 27 in which we showed that if the initial population of susceptibles $s(0)$ was less than $1/c$, we would not get an outbreak. A related way to determine whether or not an outbreak will occur is to consider the so-called *basic reproduction number* of the infection. The basic reproduction number is denoted by $\mathcal{R}_0$ and is defined to be the expected number of secondary cases produced by a single typical infection. So authors call this the "basic reproduction rate" but $\mathcal{R}_0$ is not a rate at all; in fact it is unitless.

EXERCISE 28.2. What does it mean if $\mathcal{R}_0 > 1$? If $\mathcal{R}_0 < 1$?

A formula for $\mathcal{R}_0$ is given by

$$\mathcal{R}_0 = (\text{transmissibility}) \times (\text{contact rate}) \times (\text{duration of infectiousness})$$
$$= \tau \times \bar{c} \times d$$

where τ is "infections per contact", $\bar{c}$ is "contacts per time" and d is "time per infection". Thus, we see that the product of the three has no units.

EXERCISE 28.3. Recall the SIR model in Chapter 27. Argue for why k in that notation is d^{-1} in this notation. Also, argue that b is $\tau \cdot \bar{c}$. Assuming $s(0) = 1$, conclude that we have an outbreak if $\mathcal{R}_0 > 1$.

Knowing whether or not an infection will spread in a population or not, then, comes down to knowing $\mathcal{R}_0$. There is a general technique for computing $\mathcal{R}_0$ using linear algebra [2], but here we are happy to present $\mathcal{R}_0$ for the Ross-Macdonald model, explain a little where it comes from and how it can be used.

It can be shown [8] that the basic reproduction number for the Ross-Macdonald model is given by

$$\mathcal{R}_0 = \mathcal{R}_0^{MH} \times \mathcal{R}_0^{HM}$$
$$= \tau^{MH} \bar{c}^{MH} d^{MH} \times \tau^{HM} \bar{c}^{HM} d^{HM}$$
$$= (macD_H) \times (abD_M)$$
$$= ma^2 bc D_H D_M,$$

where m is the number of mosquitoes per person. We explain the two factors in a little more detail.

By $\mathcal{R}_0^{HM}$ we mean the basic reproductive number of humans infecting mosquitoes. This number should be unitless and should be the product of transmissibility, contact rate and duration of infectiousness. Let m be the number of mosquitoes per person. Then for a single human $\mathcal{R}_0^{HM}$ should be the transmissibility (in this case c, the human to mosquito per bite transmission probability) times the contact rate (in this case ma, the number of mosquitoes per person times mosquito biting rate) times the duration of infectiousness of humans (in this case D_H). This leads us to conclude that

$$\mathcal{R}_0^{HM} = macD_H.$$

EXERCISE 28.4. Make a similar argument for why $\mathcal{R}_0^{MH}$ is equal to abD_M.

So what does this all mean? Well, recall that we want $\mathcal{R}_0 < 1$ to avoid an outbreak. Since $\mathcal{R}_0 = ma^2bcD_HD_M$, there are several ways to reduce $\mathcal{R}_0$:

- we could reduce m by reducing the number of mosquitoes in the area (e.g., via fumigation);
- we could reduce a by making it hard for a mosquito to bite (e.g., mosquito nets or mosquito repellants);
- we could shorten the mosquito life span D_M (e.g., by introducing some mosquito-specfic infection [5]); or
- we could shorten the duration of infection in a human (e.g., by treating the patient with an antimalarial drug).

EXERCISE 28.5. If we could reduce one of m, a, D_M or D_H by 50%, which should we reduce? Justify your answer.

What about vaccinations? Recall that in Chapter 27 we showed that reducing $s(0)$ by a certain amount, would lead to $\frac{di}{dt}$ always being negative and that, in turn, would lead to the infection not taking hold of the population. It turns out that if it were possible to vaccinate $H(1 - 1/\mathcal{R}_0)$ people in the population, we would prevent an outbreak. So again, the smaller the basic reproduction number is, the less likely there is to be an outbreak.

Of course there are many things this simple model does not cover. It has no spatial component and assumes that humans and mosquitoes are uniformly and immediately mixed. It has no stochastic component and assumes that all mosquitos will live long enough to bite the same number of people the same number of times. It has no seasonal component, it assumes that there are as many mosquitoes in the dry season as there are in the rainy season. It assumes that all infections by *Plasmodium* are the same, despite there being multiple strains with complicated immunity. That said, the model still does a good job helping us understand the basic structure of how malaria spreads within a population.

28.4. Problems

PROBLEM 28.6. In Exercise 28.5, you were asked to consider the sensitivity of there being an outbreak of malaria to the choice of parameters. Do a more thorough sensitivity analysis of the model to the choice of parameters. According to [4], a is between 0.01 and 0.5 day^{-1}, b is between 0.2 and 0.5, c is 0.5, m is between 0.5 and 40, ξ is between 0.05 and 0.5 day^{-1} and δ is between 0.005 and 0.05 day^{-1}.

PROBLEM 28.7. The Macdonald model [3] can be considered to be as follows:

$\frac{di_h}{dt} = abmi_m(1 - i - h) - \delta i_h$

$\frac{de_m}{dt} = aci_h(1 - e_m - i_m) - aci_h(t - \tau_m)[1 - e_m(t - \tau_m) - i_m(t - \tau_m)]e^{-\xi\tau_m} - \xi e_m$

$\frac{di_m}{dt} = aci_h(t - \tau_m)[1 - e_m(t - \tau_m) - i_m(t - \tau_m)]e^{-\xi\tau_m} - \xi e_m.$

Explain what this model tries to address that is not addressed by the model in this chapter. In particular, what are e_m, τ_m and the role of the exponential term $e^{-\xi\tau_m}$. Use this interpretation to show that $\mathcal{R}_0$ in this case is $ma^2bcD_HD_Me^{-\mu_2\tau_m}$ and explain how to use this to decrease the likelihood of there being a malaria outbreak.

Bibliography

[1] Folashade B Agusto, Sara Y Del Valle, Kbenesh W Blayneh, Calistus N Ngonghala, Maria J Goncalves, Nianpeng Li, Ruijun Zhao, and Hongfei Gong. The impact of bed-net use on malaria prevalence. *Journal of Theoretical Biology*, 320:58–65, 2013.

[2] JM Heffernan, RJ Smith, and LM Wahl. Perspectives on the basic reproductive ratio. *Journal of the Royal Society Interface*, 2(4):281–293, 2005.

[3] George Macdonald et al. *The epidemiology and control of malaria*. London, Oxford Univ. Pr., 1957.

[4] Sandip Mandal, Ram Rup Sarkar, and Somdatta Sinha. Mathematical models of malaria-a review. *Malaria Journal*, 10(1):1, 2011.

[5] Conor J McMeniman, Roxanna V Lane, Bodil N Cass, Amy WC Fong, Manpreet Sidhu, Yu-Feng Wang, and Scott L O'Neill. Stable introduction of a life-shortening wolbachia infection into the mosquito aedes aegypti. *Science*, 323(5910):141–144, 2009.

[6] Calistus N Ngonghala, Mateusz M Pluciński, Megan B Murray, Paul E Farmer, Christopher B Barrett, Donald C Keenan, and Matthew H Bonds. Poverty, disease, and the ecology of complex systems. *PLoS Biol*, 12(4):e1001827, 2014.

[7] Ronald Ross. The prevention of malaria. 1911.

[8] David L Smith, F Ellis McKenzie, Robert W Snow, and Simon I Hay. Revisiting the basic reproductive number for malaria and its implications for malaria control. *PLoS Biol*, 5(3):e42, 2007.

[9] Dorothy I Wallace, Ben S Southworth, Xun Shi, Jonathan W Chipman, and Andrew K Githeko. A comparison of five malaria transmission models: benchmark tests and implications for disease control. *Malaria Journal*, 13(1):1, 2014.

CHAPTER 29

HIV/AIDS

The epidemiological models in Chapters 27 and 28 have a couple of things in common. First, the total population in both models are constant. Second, the age of the population is not taken into account. In this chapter, as we study the population dynamics of AIDS, we present a model that changes both of these assumptions. The second change is particularly important: by considering so-called "vertical" transmission of AIDS (e.g., from a mother to a child), the model is a better representation of the reality in sub-Saharan Africa where HIV is most often spread through heterosexual sex (see [5] for a review of the barriers and promoters of vertical transmission of HIV in sub-Saharan Africa). Of course the "horizontal" transmission is still present in the model and also plays an important role.

Models for the spread of AIDS can get very complicated and the model we present in this chapter is just one of many. We have chosen to focus on this one because it is complicated enough to be interesting but simple enough to be analyzed. The original model can be found in [13]. For a general treatment of the spread of HIV/AIDS in the United States, see [8] and for a mathematical review of several models see [9]. The first mathematical models for AIDS were developed by Anderson and May [1, 2, 10].

In the context of this text, the point of this chapter is to take a paper in the literature, understand the components of the model and justify parts of their analysis.

29.1. Biological context

There are many ways in which diseases can be transmitted: our previous models have had so-called horizontal transmissions (e.g., cholera is spread by contact between, say, the saliva of an infected person and a susceptible person) and vector transmissions (e.g., malaria is transmitted via mosquitoes). In the case of HIV/AIDS, horizontal transmission can result from direct physical contact between an infected and a susceptible. Vertical transmission, on the other hand, can result from direct transfer of a disease from an infected mother to an unborn or newborn offspring. Among those diseases that can be transmitted vertically include the following: chagas, dengue fever, hepatitis B and HIV/AIDS. Vertical transmission of HIV/AIDS can occur during pregnancy, delivery or breastfeeding and is influenced by many factors, including maternal viral load and the type of delivery.

We report some data [3] on HIV/AIDS in South Africa to give you an idea of how prevalent the illness is in sub-Saharan Africa. In 2010, about a third of all pregnant women who attended public sector health-care facilities were infected. The prevalence of HIV infection among pregnant women is likely to remain high for at least the next two decades because the number of people receiving life-long antiretroviral therapy (ART) in South Africa is still increasing. In 2011, almost three out of four maternal deaths in South Africa were associated with HIV infection, as were half of all deaths of children younger than 5 years.

According to the World Health Organization, about 20% of the children infected with HIV develop AIDS in the first year of their lives, and most of them die by the age of 4 years. The remaining 80% develop symptoms of HIV/AIDS at school entry age (7-9 years) or even during adolescence.

HIV/AIDS transmission in Africa is primarily through heterosexual sex and vertical transmission (mother-to-child). Forty percent of all HIV/AIDS cases result from mother to child transmission [4]. Fewer than 300 infants in the U.S acquired HIV through vertical transmission in 1997. In sub-Saharan Africa over 2.5 million children under the age of 15 have died of AIDS. Most of these children were exposed to HIV during labor or breastfeeding. HIV/AIDS is globally a serious threat to future development.

The impact of vertical transmission of HIV/AIDS has been felt mainly in Africa, where the level of literacy is low, the poverty level is very high and the quality of health services is generally very poor. The current recommendations to manage vertical transmission is either dual therapy with nevirapine and azidothymidine from the 14th week of pregnancy onward or highly active antiretroviral therapy. Other studies involving treatment of HIV-infected children have demonstrated further that effective treatment for these children can prolong their survival and significantly improve the quality of their lives. The current antiretroviral drugs (ARV) are known to be effective in lowering viral loads, and the infected children may as a result reach adulthood and become sexually active.

Naresh *et al.* [13] developed a mathematical model on the dynamics of HIV/AIDS epidemic with vertical transmission and then [12] proposed an HIV/AIDS model with vertical transmission and the period of sexual maturity of infected newborns which is incorporated in the model as a time delay. The model below is based on [13].

29.2. The model

We consider a population of size $N(t)$ at time t with constant inflow of susceptibles at a rate of Q_0. The population size $N(t)$ is divided into four subclasses: susceptibles $S(t)$, infecteds $I(t)$ (also assumed to be infectious), pre-AIDS patients $P(t)$ and AIDS patients $A(t)$. Every person in population is assumed to have a natural (i.e., not induced by the disease) mortality rate of d. In the model, we assume that the susceptibles become HIV infected via sexual contacts with infected and this may also lead to the birth of infected children. It is assumed that a fraction of new born children are infected at birth and hence are directly recruited into the infective class with a rate $(1 - \varepsilon)\theta$ and others die effectively at birth $(0 \leq \varepsilon \leq 1)$. We do not consider direct recruitment of other infected persons but by vertical transmission only. The interaction between susceptibles and infecteds is assumed to be as in Lotka-Volterra. It is also assumed that some of the infecteds move to join pre-AIDS class, depending on the viral counts, with a rate $\sigma\delta$ and then proceed with a rate μ to develop full blown AIDS while others with serious infection directly join the AIDS class with a rate $(1 - \sigma)\delta$ where $0 \leq \sigma \leq 1$.

EXERCISE 29.1. Draw a box model that represents this system.

In view of the above, spread of the disease is assumed to be governed by the following system of differential equations:

$$\frac{dS}{dt} = Q_0 - \frac{\beta_1 cSI}{N} - \frac{\beta_2 cSP}{N} - \frac{\beta_3 cSA}{N} - dS,$$

$$\frac{dI}{dt} = \frac{\beta_1 cSI}{N} + \frac{\beta_2 cSP}{N} + \frac{\beta_3 cSA}{N} - (\delta + d)I + (1 - \varepsilon)\theta(I + P + A),$$

$$\frac{dP}{dt} = \sigma\delta I - (\mu + d)P,$$

$$\frac{dA}{dt} = (1 - \sigma)\delta I + \mu P - (\alpha + d)A,$$

$$S(0) = S_0, I(0) = I_0, P(0) = P_0 \text{ and } A(0) = A_0,$$

where c is the average number of sexual partners per unit time; δ is the rate of movement from infectious class, so that $\frac{1}{\delta}$ denotes the average incubation period; β_i $(i = 1, 2, 3)$ are the contact rates of susceptibles with infectives, pre-AIDS and AIDS patients respectively and α is the disease induced death rate due to AIDS. It is assumed that all the dependent variables and parameters of the model are non-negative.

To simplify the model, the authors of [13] assume that the AIDS patients and those in pre-AIDS class are sexually inactive as they are isolated and hence are not producing children.

EXERCISE 29.2. Which terms should be set to zero in order to represent this assumption?

They further assume that AIDS patients and those in the pre-AIDS class also do not contribute to vertical transmission.

EXERCISE 29.3. Which terms should be set to zero in order to represent this assumption?

The authors claim that the assumptions are valid in developed countries but are less true in under-developed countries; we leave it to the reader to develop a model better-suited to under-developed countries.

In the end, the authors present this system of differential equations:

(29.1)
$$\frac{dN}{dt} = Q_0 - dN - \alpha A + (1 - \varepsilon)\theta I$$

$$\frac{dI}{dt} = \frac{\beta_1 c(N - I - P - A)I}{N} - (\delta + d)I + (1 - \varepsilon)\theta I$$

$$\frac{dP}{dt} = \sigma\delta I - (\mu + d)P$$

$$\frac{dA}{dt} = (1 - \sigma)\delta I + \mu P - (\alpha + d)A$$

$$N(0) = N_0, I(0) = I_0, P(0) = P_0 \text{ and } A(0) = A_0.$$

EXERCISE 29.4. How can you deduce (29.1) from the earlier model and the above assumptions? Note: the first equation is for $\frac{dN}{dt}$.

29.3. Implications of the model

We present the implications of the model as a series of exercises.

EXERCISE 29.5. Suppose we want to study $I(t)$ when t is small: in this case $S \approx N$ and $A \approx P \approx 0$. Based on this assumption, show that in the early stages, the infecteds have exponential growth. In particular, show

$$I(t) = I_0 e^{(\beta_1 c + (1 - \varepsilon)\theta - (\delta + d))t}$$

for small t.

EXERCISE 29.6. It can be shown that the basic reproduction number is $\mathcal{R}_0 = \frac{\beta_1 c + (1-\varepsilon)\theta}{\delta + d}$. Justify that this number is unitless and can be interpreted in the standard way (as the sum of two basic reproduction numbers).

EXERCISE 29.7. Show that, for small t,

$$I(t) = I_0 e^{\frac{\mathcal{R}_0 - 1}{T} t},$$

where $T = \frac{1}{\delta + d}$ is the average duration of infection.

EXERCISE 29.8. What is the doubling time during the early stage of the epidemic? How do you interpret your formula if $\mathcal{R}_0 < 1$?

EXERCISE 29.9. Show that, in the absence of infection (therefore $P = I = A = 0$), the population stabilizes at $\frac{Q_0}{d}$. This is called the *disease-free equilibrium*.

EXERCISE 29.10. In order to avoid an outbreak we need $\mathcal{R}_0 < 1$. Using the formula for $\mathcal{R}_0$ derived above for this model, which rates would need to be reduced in order to decrease $\mathcal{R}_0$?

EXERCISE 29.11. Plot solutions to the model in Sage and determine whether of 50% reduction in horizontal transmission has a bigger effect on the infecteds population or a 50% reduction in vertical transmission.

29.4. Problems

PROBLEM 29.12. In the 1990s, the transmission of HIV between married couples in Uganda with different HIV statuses was studied [6]. It was found that the transmission probability was, on average, 0.0011 per coital act and that there was an average of 9 coital acts per month. Almost all the couples in the study did not use condoms. Also in the 1990s, it was found that in Uganda about 26% of babies contracted HIV from their mothers during childbirth [7, 11]. Incorporate these values into the model and determine what would happen with a reasonable increase in condom use or a reasonable increase in the protection of infants (also, what is "reasonable").

PROBLEM 29.13. In the model above, the treatment of infecteds, pre-AIDS or AIDS patients is not taken into account. How would you change the model to incorporate treatment? Compare your results to [14].

Bibliography

[1] RM Anderson, GF Medley, RM May, and AM Johnson. A preliminary study of the transmission dynamics of the human immunodeficiency virus (HIV), the causative agent of AIDS. *Mathematical Medicine and Biology*, 3(4):229–263, 1986.

[2] Roy M Anderson. The role of mathematical models in the study of HIV transmission and the epidemiology of AIDS. *JAIDS Journal of Acquired Immune Deficiency Syndromes*, 1(3):241–256, 1988.

[3] Peter Barron, Yogan Pillay, Tanya Doherty, Gayle Sherman, Debra Jackson, Sanjana Bhardwaj, Precious Robinson, and Ameena Goga. Eliminating mother-to-child HIV transmission in South Africa. *Bulletin of the World Health Organization*, 91(1):70–74, 2013.

[4] Kevin M De Cock, Mary Glenn Fowler, Eric Mercier, Isabelle de Vincenzi, Joseph Saba, Elizabeth Hoff, David J Alnwick, Martha Rogers, and Nathan Shaffer. Prevention of mother-to-child HIV transmission in resource-poor countries: translating research into policy and practice. *Jama*, 283(9).1175–1182, 2000.

[5] Annabelle Gourlay, Isolde Birdthistle, Gitau Mburu, Kate Iorpenda, and Alison Wringe. Barriers and facilitating factors to the uptake of antiretroviral drugs for prevention of mother-to-child transmission of hiv in sub-saharan africa: a systematic review. *Journal of the International AIDS Society*, 16 (1), 2013.

[6] Ronald H Gray et al. Probability of hiv-1 transmission per coital act in monogamous, heterosexual, hiv-1-discordant couples in rakai, uganda. *The Lancet*, 357(9263):1149–1153, 2001.

[7] Laura A Guay et al. Intrapartum and neonatal single-dose nevirapine compared with zidovudine for prevention of mother-to-child transmission of HIV-1 in Kampala, Uganda: HIVNET 012 randomised trial. *The Lancet*, 354(9181): 795–802, 1999.

[8] Herbert W Hethcote and James W Van Ark. *Modeling HIV transmission and AIDS in the United States*, volume 95. Springer Science & Business Media, 2013.

[9] Valerie Isham. Mathematical modelling of the transmission dynamics of HIV infection and AIDS: A review. *Journal of the Royal Statistical Society. Series A (Statistics in Society)*, 151(1):5–49, 1988.

[10] Robert M May and Roy M Anderson. COMMENTARY: Transmission dynamics of HIV infection. *Nature*, 326:137, 1987.

[11] Philippe Msellati, Marie-Louise Newell, and Francois Dabis. Rates of mother-to-child transmission of HIV-1 in Africa, America and Europe: results from 13 perinatal studies. *Journal of Acquired Immune Deficiency Syndromes*, 8: 506–510, 1995.

[12] Ram Naresh and Dileep Sharma. An HIV/AIDS model with vertical transmission and time delay. *World Journal of Modelling and Simulation*, 7(3):230–240, 2011.

[13] Ram Naresh, Agraj Tripathi, and Sandip Omar. Modelling the spread of AIDS epidemic with vertical transmission. *Applied Mathematics and Computation*, 178(2):262–272, 2006.

[14] Abdallah S Waziri, Estomih S Massawe, and Oluwole Daniel Makinde. Mathematical modelling of HIV/AIDS dynamics with treatment and vertical transmission. *Applied mathematics*, 2(3):77–89, 2012.

CHAPTER 30

Projects for Infectious Disease Models

At this point, we hope that the reader will have developed the confidence to ask their own questions and develop their own models. So, instead of listing some projects for you to consider, we share with you some abstracts from papers written by students in a recent class that used this text. The assignment was to find an existing model and tweak it a little and then to analyze the model. We invite the reader to be inspired by these examples to set out in yet another direction or, at least, to maybe tweak one of these tweaks.

(1) Marie Confreda wrote a paper entitled "SIRC Model with Vaccines and Influenza A" and whose abstract read:

> The spread of diseases can be modeled for a population. In this paper the spread of Influenza A is modeled using a SIRC model. In this model there are people who are susceptible (S), infected (I), recovered (R), and cross-immune (C) to Influenza A. The cross-immune are those that recovered after being infected by different strains of the same viral subtype in the past years. I altered this model by introducing vaccinations, and this change will make the model more accurate to a realistic scenario, since people receive vaccinations for the flu. The model suggests that with vaccinations the number of people who are infected will decrease compared to the model without vaccinations. These findings are consistent with data, which suggests that flu vaccinations should decrease the amount of people who are infected with the flu. It is also shown that the amount of people who are vaccinated and the effectiveness of the vaccine are the most important parameters with vaccinations.

This paper was an improvement to [2]. An assumption of this paper is that there is only one circulating strain of influenza A. How might one model more the system if there were more than one strain circulating?

(2) Julia Laibach wrote a paper entitled "Comparison and Analysis of Marine and Freshwater Bacteriophage Infections" and whose abstract was

> The infection of marine and freshwater bacteria by lytic bacteriophages is described by a mathematical model used to compare the behaviors of the two infections. The factors pH and irradiance indirectly affect the efficiency of bacteriophage infections by directly influencing the behavior of viral particles. As most bacterial cell walls are negatively charged, a bacteriophage with a more positive charge will have a higher amount of contact with negatively charged bacteria. The charge of a bacteriophage is influenced by the pH of its external environment. As pH

241

decreases, the bacteriophage becomes more positively charged. Irradiance directly affects the survival rate of photoautotrophic bacteria. An environment that absorbs more sunlight will provide less light for bacterial populations. Marine environments are characterized by clearer, more basic water conditions while freshwater environments are characterized by more acidic water with a higher level of light reflecting and absorbing particles. We proved that the general behavior of marine and freshwater bacteriophage infections is similar, but freshwater bacteriophages are more successful in initiating a successful infection.

This paper was based on [1]. This paper focused on how in pH and irradiance in fresh- versus saltwater affected bacteriophage infections. One could consider other factors such as temperature and oxygen levels.

(3) Ben Barrett wrote an article entitled "Inclusion of a Second Latency Term in a SARS Double Epidemic Model" and whose abstract read

Severe Acute Respiratory Syndrome (SARS) arose in the Guangdong province of China in November 2002 and then spread to the rest of the world, resulting in 774 deaths before the epidemic was eradicated in 2003. SARS is caused by a coronavirus. It has been proposed that while SARS was spreading a second coronavirus, termed Virus B, was also active that granted people immunity from SARS if they were infected with Virus B first. Ng, Turinici, & Danchin (2003) first suggest this hypothesis, and this paper builds upon their original assumptions by including a latency period for both SARS and Virus B. It is discovered that a model including a latency period for both viruses matches actual Hong Kong SARS data just as well as a model with no latency period for Virus B. This suggests that it is quite possible for Virus B to have a latency period, and if it does, this latency period is far shorter than that of SARS. This developed model can be applied to other coronaviruses, such as MERS, to aid in predicting virus spread and steering health policy decisions.

The paper was based on [3]. A further refinement of the model would be to apply it to Middle East respiratory syndrome corona virus.

Bibliography

[1] Edoardo Beretta and Yang Kuang. Modeling and analysis of a marine bacteriophage infection. *Mathematical Biosciences*, 149(1):57–76, 1998.

[2] Renato Casagrandi, Luca Bolzoni, Simon A Levin, and Viggo Andreasen. The sirc model and influenza a. *Mathematical Biosciences*, 200(2):152–169, 2006.

[3] Tuen Wai Ng, Gabriel Turinici, and Antoine Danchin. A double epidemic model for the sars propagation. *BMC Infectious Diseases*, 3(1):1, 2003.

CHAPTER 31

Classroom Tested Projects

31.1. Population modeling

Zebra mussels. The *Dreissena polymorpha*, or zebra mussel, is thought to have made its way to North America by latching onto a voyaging European ship in 1988. Since its introduction to the Great Lakes, the zebra mollusk has proliferated to a population that is difficult to estimate. Model the growth of zebra mussels and estimate current and future populations of it.

Possible references:

- Amiad Filtration Systems (1997). "Effective removal of zebra mussels with Amiad screen technology."
- "Great Lakes Fact Sheet." EPA. Environmental Protection Agency, n.d. Web. 9 May 2014.
- Gurevitch, J., Padilla, D.K. (2004). "Are invasive species a major cause of extinctions?" *Trends in Ecology & Evolution*, Vol. 19 (9).
- Melius, Tom. "North Texas Zebra Mussel Barrier." U.S. Department of the Interior. N.p., n.d. Web. 10 May 2014.
- Nalepa, T. F. and Donald W. Schloesser. "Quagga and Zebra Mussels: Biology, Impacts, and Control." 2nd ed. Boca Raton: CRC, 2014. Print.
- National Atlas (2013). Zebra Mussels.
- Tammi, Karin A. "Zebra Mussel: An Unwelcome Visitor." Rhode Island Sea Grant, n.d. Web. 9 May 2014.
- University of Wisconsin Sea Grant Institute, (2013). *Zebra Mussels (Dreissena polymorpha)*.
- Waller, D.L., Rach, J.J., Cope, W.C. and Marking, L.L. (1993). "Toxicity of candidate molluscicides to zebra mussels (*Dreissena polymorpha*) and selected nontarget organisms." *J. Great Lakes Res.* 19(4).

31.2. Drug modeling

Filiarisis. Here you will investigate the relative merits of three dosage strategies for Diethylcarbamazine (DEC), a treatment for filiarisis. You need the chapter on DEC from the Handbook of Drugs for Tropical Parasitic Infections, which includes a description of the amount usually given and how they are divided into several doses per day, given orally. Basic information about the disease can be found on Wikipedia and basic information about the medication (e.g., the medicine is liquid and the time to mix it and send it to the GI tract is negligible) can be found on the CDC website. The CDC recommends a 1-12 day treatment and 6 mg/kg bodyweight per dose.

This will be a basic drug study: Assume that the drug follows a standard GI tract/blood concentration compartment model. Develop and parametrize this model using the information in

- Hellgren, Urban, Orjan Ericsson, Yakoub AdenAbdi and Lars L. Gustafsson. *Handbook of drugs for tropical parasitic infections*. CRC Press, 1995.

Use Sage to compare what happens over 20 days with a single dose, 2 doses 6 days apart, 3 doses 4 days apart, 4 doses 3 days apart, and daily doses for 12 days. Keep a data set with the following values for each run:

- F, V, k_a, k_{el}, Run#, Day, Time, Length of run, starting X (mass of drug), starting C (drug concentration), ending X, ending C, max C, min C. You will turn in your data tables, along with plots of a few runs.

After the first dose, what are the maximum and minimum blood concentrations over this period? Because the constants of absorption and elimination fall in a range, you will need to test the extreme values for these.

(1) Research question: How do the maximum and minimum blood concentrations of this drug depend on the dosage schedule?
(2) Research question: Read up a bit on this drug. Based on your reading and your computational experiments, which dosage regimen do you recommend?
(3) Research question: How would impaired kidneys (or liver, see reference 9 at the end of the Handbook) affect the dosage regime you have designed? Use your model to suggest an alternative for such individuals.

Bortezomid. Based on the paper cited below, you and your partner will develop a two-compartment model for IV bolus administration of Bortezomid. Use Sage to identify parameters that give a model that matches the data in Papandreou *et al.* Research questions:

(1) What box model and equations best represent the dynamics of this drug in the human body? What constants must be identified from data?
(2) Based on the description given in the paper, how would you set initial conditions for the doses described in Figure 1? Notice the log scale.
(3) Using the dosage 1.45, find constants that give Big Green runs that match the data from Figure 1. Note the log scale. You should receive correct data in a spreadsheet from math 27.
(4) Does the AUC given in Table 6 for a single dose agree with your model? Figure out how to compute it and show the run.
(5) The paper describes giving a second dose on day 8. How much is left from the first dose, in blood and soft tissue, at that time, according to your model?
(6) Simulate the second dose. Does anything different happen?
(7) Think about the dosage regime in comparison to potential side effects. Read the paper. Think about whether 8 days is a realistic interval for this therapy.
(8) Simulate the actual experiment, one cycle (days 1,8,15,22, for four total injections).
(9) How much is in the blood/soft tissue after the day 22 injection? How long does it take for the system to be almost empty? What are the

concentrations on day 35 when the next cycle starts? Does it look ok or would you wait longer?

(10) At what point in the cycle do you think toxicity would be the highest? What would you tell a patient to expect?

The paper referred to above is

- Papandreou, Christos N. et al. "Phase I trial of the proteasome inhibitor bortezomib in patients with advanced solid tumors with observations in androgen-independent prostate cancer." *Journal of Clinical Oncology* 22.11 (2004): 2108–2121.

2-DG versus ABT. You are hired by a pharmaceutical company to model the likely pharmacokinetics of two drugs used in cancer therapy, 2-DG and ABT. ABT is an agent that induces apoptosis (cell death) in some types of tumors. According to one research group, 2-DG (a glucose derivative) increases the effectiveness of ABT, allowing it to target more types of cancer cells without harming healthy cells. The researchers in that group injected 2-DG three hours in advance of injecting the ABT. In order to move this research to human trials, your company needs to know a few things.

(1) What are the basic pharmacokinetic models for these two drugs? Figure out all relevant constants for the models (from the research papers I have given to you). These are F, V, k_a, k_{el}, half-life, time to peak, and peak concentration (which depends on the dosage given, which you need to include). Also AUC, as the papers you a reading use that parameter.

(2) Get your models working on Sage. For experimental purposes we will take the dosages of ABT as 100 mg/kg (100 mg per kilogram of body weight). We will take 45 mg/kg as the recommended dose of 2-DG. Both are taken orally as a liquid that we can assume is immediately available for absorption in the GI tract. You have to make an assumption about body weight for your virtual patient.

(3) The mouse study (Yamaguchi *et al.*, see below) gives a dose of ABT once per day, pretreated with 2-DG 3 hours in advance. This is the general plan your company wants to follow, only with humans and orally.

(4) So you need to design a treatment schedule that gives both 2-DG and ABT once per day. You want the 2-DG to be within some tolerance (say 80 percent) of peak value for at least three hours prior to administering the ABT. ABT is stable in solution for only about 4 hours. You want the ABT to be stable and high (say within 80 percent of peak) for as long as possible after the three hours of high 2-DG. Only the first four hours of the ABT dose is likely to be of any clinical use.

(5) So you are trying to answer this question: What is the greatest length of time I can keep ABT within 80 percent of peak, AND have 2-DG within 80 percent of peak for the prior 3 hours, AND have all of this happen within the first 4 hours of receiving the ABT dose? And how did I do it (with what dosage protocol)?

(6) Is there a problem with multiple doses? Try three days in a row.

(7) Having solved this problem you should now go back and look at the pharmacokinetic constants and change each one up and down 10 percent to see

what happens to your answer. After all, these constants are measurements that also have some error.

The paper referred to above:

- Yamaguchi, Ryuji et al. "Efficient elimination of cancer cells by deoxy-glucose-ABT-263/737 combination therapy." *PLoS One* 6.9 (2011): e24102.

31.3. Predator-Prey modeling

Skuas and penguins. Model the effects of the predation rate of skuas on the population of penguins that reach adulthood. Using data from the book *Skua and Penguin: Predator and Prey* by Euan Young, formulate equations to represent each stage of a penguins' development and the predation upon them. Two eggs are laid per breeding pair and hatch two weeks apart. To what extent does the two-egg strategy increase the survival rate?

- Young, E. (1994). *Skua and Penguin: Predator and Prey.* Cambridge: Cambridge University Press.

Strawberry poison-dart frog. The *Dendrobates Pumilio* or Strawberry poison-dart frog exhibits vast differences in coloration between individuals as a result of genetic differences within the population. Construct a model of the flow of genotypes in the population as a result of mate selection by female frogs.

- Richards-Zawacki, C. L., Wang, I. J. and Summers, K. (2012), "Mate choice and the genetic basis for colour variation in a polymorphic dart frog: inferences from a wild pedigree." *Molecular Ecology*, 21:3879–3892.

Honey bees. Western Honeybees, *Apis mellifera*, are susceptible to the Varroa destructor parasite, which can cause a significant decline in honeybee populations. Construct a model that examines the effect of Varroa mites infiltrating a honeybee colony and the effect on the forager and hive bee subpopulations.

A list of references:

- Eberl, Hermann J., Mallory R. Frederick and Peter G. Kevan. "Importance of Brood Maintenance Terms in Simple Models of the Honeybee-Varroa Destructor- Acute Bee Paralysis Virus Complex." *Electronic Journal of Differential Equations*, Conf. 19 (2010): 85–98.
- The Food & Environment Research Agency. "Managing Varroa." 2013. Digital file.
- Khoury, David S., Mary R. Myerscough and Andrew B. Barron. "A quantitative model of honey bee colony population dynamics." *PLoS ONE* 6.4 (2011) e18491.
- Sumpter, D.J.T. and S.J. Martin. "The dynamics of virus epidemics in Varroa-infested honey bee colonies." *Journal of Animal Ecology* (2004): 51–63.

Dolphins and porpoises. Build a model of the relationship between harbor porpoises and bottlenose dolphins (which both compete with porpoises and kill them) in the Moray Firth, and analyze this model to determine possible prey abundance necessary for dolphins tpstop killing porpoises over competition.

Some useful references:

- Clark, Nicola. "The Spatial and Temporal Distribution of the Harbour Porpoise (*Phocoena Phocoena*) in the Southern Outer Moray Firth, NE Scotland." Masters of Science Thesis, *Marine Mammal Science* (2005): 1102.
- Ólafsdóttir, D., Víkingsson, G. A., Halldórsson, S. D. and Sigurjónsson, J. 2002. "Growth and reproduction in harbour porpoises (*Phocoena phocoena*) in Icelandic waters." *NAMMCO Sci. Pub.* 5:195210.
- Spitz, J., Rousseau, Y. and Ridoux, V. "Diet Overlap between Harbour Porpoise and Bottlenose Dolphin: An Argument in Favour of Interference Competition for Food?" *Estuarine, Coastal and Shelf Science* 70.12 (2006).
- Thompson, P. M., Lusseau, D., Corkrey, R. and Hammond, P. S. (2004). "Moray Firth bottlenose dolphin monitoring strategy options." Scottish Natural Heritage Commissioned Report No. 079 (ROAME No. F02AA409).

Wolf spiders. How does mortality from cannibalism affects the population regulation of the cursorial wolf spider *Schizocosa ocreata*, which can be found in the leaf litter of forests across the eastern United States? The high densities of recently dispersed young *S. ocreata* decreased dramatically by time autumn. Make a model of the experiments they conducted and attempt to replicate their results.

Some references:

- Wagner, James D. and David H. Wise. "Cannibalism regulates densities of young wolf spiders: evidence from field and laboratory experiments." *Ecology* (1996): 639–652.
- Wise, David H. and James D. Wagner. "Evidence of exploitative competition among young stages of the wolf spider *Schizocosa ocreata*." *Oecologia* 91.1 (1992): 7–13.
- Buddle, Christopher M. "Life History of *Pardosa Moesta* and *Pardosa Mackenziana* (*Araneae, Lycosidae*) in Central Alberta, Canada". *The Journal of Arachnology* 28.3 (2000): 319–328.
- Uetz, G. W., Papke, R. and Kilinic, B. (2002). "Influence of feeding regime on body size, body condition and a male secondary sexual character in *Schizocosa ocreata* wolf spiders (*Araneae, Lycosidae*): condition-dependence in a visual signaling trait". *Journal of Arachnology* 30, 461–469.

31.4. Infectious disease modeling

Influenza. Model an outbreak of influenza both absent from and in the presence of vaccination. Include both the traditional vaccine and the more novel FluMist inhalable version. These give different levels of protection. Some references:

- "Key Facts About Seasonal Flu Vaccine." Seasonal Influenza (Flu). Centers for Disease Control and Prevention, 21 Sept. 2011.
- Qiu, Zhipeng and Zhilan Feng. "Transmission Dynamics of an Influenza Model with Vaccination and Antiviral Treatment." *Bulletin of Mathematical Biology* 72 (2010): 1–33.
- Roos, Robert. "Shot Beats Nasal Spray in Adult Seasonal Flu Vaccine Trial." *CIDRAP.* 29 Sept. 2009.

Sensitivity analysis on influenza outbreaks. Conduct a sensitivity analysis, to identify the variables that most affect the time between peaks of influenza outbreaks to see which areas of influenza control should be adamantly targeted. Some references:

- Holloran, M. Elizabeth, Frederick G. Hayden, Yang Yang, Ira M. Longini, Jr. and Arnold S. Monto. "Antiviral Effects on Influenza Viral Transmission and Pathogenicity: Observations from Household-based Trials". *American Journal of Epidemiology.* 165(2):212–221 (2006).
- Carrat, Fabrice, Elisabeth Vergu, Neil M. Ferguson, Magali Lemaitre, Simon Cauchemez, Steve Leach and Alain-Jacques Valleron. "Time Lines of Infection and Disease in Human Inuenza: A Review of Volunteer Challenge Studies". *American Journal of Epidemiology.* 167(7):775-785 (2007).
- Mathews, John D., Christopher T. McCaw, Jodie McVernon, Emma S. McBryde and James M. McCaw. "A Biological Model for Influenza Transmission: Pandemic Planning Implications of Asymptomatic Infection and Immunity". *PLoS ONE.* 2(11):e1220. doi:10.1371/journal.pone.0001220.
- Ferguson, Neil M., Derek A. T. Cummings, Christopher Fraser, James C. Cajka, Philip C. Cooley and Donald S. Burke. "Strategies For Mitigating an Influenza Pandemic". *Nature.* 442(27):448-452 (2006).

Schistosomiasis. Evaluate the transmission dynamics of schistosomiasis. What is the best way to control this disease in a particular province of China via two different intervention methods — Praziquantel and Niclosamide? Some references:

- Allen, E.J. and H.D. Victory, Jr. "Modelling and simulation of a schistosomiasis infection with biological control", *Acta Trop.* 87 (2003), pp. 257–267.
- Barbour, AD (1996). "Modeling the transmission of schistosomiasis: an introductory view". *American Journal of Tropical Medicine and Hygiene* 55, 135–143.
- Chiyaka, E.T. and Garira, W. (2009). "Mathematical analysis of the transmission dynamics of schistosomiasis in the human-snail hosts". *J Biol Syst* 17: 397–423.
- Flisser, A. and D.J. McLaren (1989). "Effect of praziquantel treatment on lung stage larvae of Schistosoma mansoni in vivo". *Parasitology*, 98: 203–211.
- Gryseels B. "The relevance of schistosomiasis for public health". *Tropical and medical parasitology.* 1989, 40: 134–142.
- Gu, XG and Kang JX. (1999). "Progress of schistosomiasis control in mountainous regions of China". *Chinese Journal of Schistosomiasis Control* 11, 261–262.
- Hu, Z. (1990). "Freshwater Snails in the Changsha Region China". *Acta Scientiarum Naturalium Universitatis Normalis Hunanensis*, 13(3), 264–267.

Ebola. Using a SEIR model developed by G. Chowell *et al.* and the 1995 Congo outbreak data, model potential infection rates that could lead to whole populations becoming ill. How do parameters affect duration of infection and death rates?

- Chowell, G., N.W. Hengartner, C. Castillo-Chavez, P.W. Fenimore and J.M. Hyman. "The Basic Reproductive Number of Ebola and the Effects of Public Health Measures: the Cases of Congo and Uganda." *J. Theoretical Biology* 229.1 (2004): 119–126.
- Khan, A.S. et al. "The Reemergence of Ebola Hemorrhagic Fever, Democratic Republic of the Congo, 1995." *Journal of Infectious Diseases.* 179. Supplement 1 (1999): S76–S86.
- "Known Cases and Outbreaks of Ebola Hemorrhagic Fever, in Chronological Order." Centers for Disease Control and Prevention. 12 Oct. 2011.
- Mackinnon, Margaret J. and Andrew F. Read. "Virulence in Malaria: an Evolutionary Viewpoint." *Phil. Trans. R. Soc. Lond.* B 359 (2004): 965–986.
- Mahanty, Siddhartha, Manisha Gupta, Jason Paragas, Mike Bray, Rafi Ahmed and Pierre E. Rollin. "Protection from Lethal Infection is determined by Innate Immune Responses in a Mouse Model of Ebola Virus Infection." *Virology* 312 (2003): 415–424.
- Sadek, Ramses F., Ali S. Khan, Gary Stevens, C.J. Peters and Thomas G. Ksiazek. "Ebola Hemorrhagic Fever, Democratic Republic of the Congo, 1995: Determinants of Survival" *J. Infectious Diseases* 179.S1 (1997): S24–S27.

Tasmanian devil tumor facial disease. Managing Tasmanian devil facial tumor disease (DFTD) through selective culling is difficult, but a diagnostic test capable of detecting the disease could help significantly. Model the early and late stage culling of individuals to determine if the development of such a diagnostic test is worthwhile. Some references:

- Beeton, Nick and Hamish Mccallum. "Models Predict That Culling Is Not a Feasible Strategy to Prevent Extinction of Tasmanian devils from Facial Tumour Disease." *Journal of Applied Ecology* 48.6 (2011): 1315323.
- "Captive Population." Save the Tasmanian Devil Program. Tasmanian Government Department of Primary Industries, Parks, Water and Environment.
- Guiler, E. R. (1970). "Observations on the Tasmanian Devil, *Sarcophilus harrisii (Marsupialia : Dasyuridae)* II. Reproduction, breeding, and growth of pouch young". *Australian Journal of Zoology*, 18, 4962.
- Lachish, Shelley, Jones, Menna and Mccallum, Hamish. "The Impact of Disease on the Survival and Population Growth Rate of the Tasmanian Devil." *Journal of Animal Ecology.* British Ecology Society, 2007.
- McCallum, H., Jones, M., Hawkins, C., Hamede, R., Lachish, S., Sinn, D. L., Beeton, N. and Lazenby, B. (2009). "Transmission dynamics of Tasmanian devil facial tumor disease may lead to disease induced extinction". *Ecology*, 90, 3379–3392.

Index

Printed in the United States
By Bookmasters